Managing Packaging Design for Sustainable Development

Managing Packaging Design for Sustainable Development

A Compass for Strategic Directions

Daniel Hellström and Annika Olsson

with contributions from
Fredrik Nilsson

This edition first published 2017 © 2017 by John Wiley & Sons, Ltd

Registered Office
John Wiley & Sons, Ltd, The Atrium, Southern Gate, Chichester, West Sussex, PO19 8SQ, UK

Editorial Offices
9600 Garsington Road, Oxford, OX4 2DQ, UK
The Atrium, Southern Gate, Chichester, West Sussex, PO19 8SQ, UK
111 River Street, Hoboken, NJ 07030-5774, USA

For details of our global editorial offices, for customer services and for information about
how to apply for permission to reuse the copyright material in this book please see our website at
www.wiley.com/wiley-blackwell.

The right of Daniel Hellström and Annika Olsson to be identified as the authors of this work has
been asserted in accordance with the UK Copyright, Designs and Patents Act 1988.

Library of Congress Cataloging-in-Publication data applied for:

ISBN: 9781119150930

A catalogue record for this book is available from the British Library.

Wiley also publishes its books in a variety of electronic formats. Some content that appears in print
may not be available in electronic books.

Cover image: Erik Andersson

Set in 10/12pt Warnock by SPi Global, Pondicherry, India

Printed and bound in Malaysia by Vivar Printing Sdn Bhd

10 9 8 7 6 5 4 3 2 1

Contents

About the Authors

Daniel Hellström is an Associate Professor in Packaging Logistics, www.plog. lth.se, at the Department of Design Sciences, Lund University, Sweden. He earned his PhD from Lund University in 2007. His research has appeared in journals including *Packaging Technology and Science, International Journal of Physical Distribution & Logistics Management, Transportation Research Part E: Logistics and Transportation Review* and *Journal of Business Research*. His research is characterized as multidisciplinary and is closely related to industrial practice. Specifically, he enjoys research regarding technology, retail, and logistics and supply chain management. He has been active in establishing the research platform ReLog (Retail Logistics), www.relog.lth.se, and the Centre for Retail Research at Lund University, www.handel.lu.se. He participates in several national and international research projects and educational programmes. He enjoys teaching Master's students and executives, and supervising PhD students.

Annika Olsson holds the Bo Rydin Professorship in Packaging Logistics, www. plog.lth.se, at the Department of Design Sciences, Lund University. She earned her PhD from Lund University in 2006. Her research is mainly on user-oriented packaging innovation and packaging development for sustainable development in supply chains and for society. Her particular research focus is on food and packaging supply chains, which she carries out in close collaboration with the related industries. Professor Olsson has had more than 15 years experience of working in the Swedish food and packaging industry. She is active in the management of the research platform ReLog (Retail Logistics), www.relog.lth.se, and the Centre for Retail Research at Lund University, www.handel.lu.se. Her teaching activities are related to packaging technology and development. Professor Olsson supervises Master's and PhD students in the areas of packaging development, packaging innovation and packaging logistics. She has published research in journals including *Packaging Technology and Science, Journal of Cleaner Production, Technovation, British Food Journal, International Journal of Logistics Research and Applications* and *The International Review of Retail, Distribution and Consumer Research*.

Fredrik Nilsson is Professor in Packaging Logistics, www.plog.lth.se, at the Department of Design Sciences, Lund University, Sweden. He is also Professor Extraordinary at Stellenbosch University, South Africa, where he has established research and education in the field of packaging logistics. His research areas are complexity thinking and theory, with current projects in health care, packaging and consumer goods supply chains. In close cooperation with a large number of partners from industry and academia, he is now dedicated to food waste issues and trying to address this major problem with new packaging solutions that integrate innovative thinking with mobile technologies. He has published research in journals including *International Journal of Operations and Production Management*, *International Journal of Logistics Management*, *International Journal of Physical Distribution & Logistics Management*, *International Journal of Business Logistics* and *International Journal of Retail & Distribution Management*.

Preface

This book is about packaging design for sustainable development, the kind of design that can make our lives friendlier, our planet greener and our businesses richer.

It is necessary and obvious that we need to move towards a more sustainable society, as we see more pollution in our oceans, more waste in our streets, more landfills and the tremendous waste of essential resources such as food on our planet. Despite the negative effects that packaging might have on our world, we need to see the other side of the coin and ask: What can packaging do to contribute to sustainable development? Interestingly enough, there is research evidence indicating that packaging design initiatives have a major impact on sustainable development.

Packaging design is a powerful vehicle for change in making the transition to a more sustainable society. What is missing is a compass that can guide practitioners in the right direction. This is particularly so in the field of packaging, where the routes you take may contradict rather than contribute to sustainable development. *Managing Packaging Design for Sustainable Development* presents a compass for you to find the path to get there. With the compass we encourage you to go off-road, to develop and innovate, and to remake the packaging design solution that previously was best practice. In a world of continuous change, technology, people and organizations keep changing the routes we take to attain sustainable development. In this world, a compass is more important than ever before.

The intention of this book goes beyond presenting a compass. The overall ambition is to bring order out of chaos in a multidisciplinary field where misconceptions and contradictory views are more dominant than the coherence and recognition of its importance. *Managing Packaging Design for Sustainable Development – A Compass for Strategic Directions* is strongly grounded in the concept that the book as a whole has a far more important story to tell than presenting every little detail. Even though reading the book makes you zoom in on packaging design, its aim is to empower you to zoom out and gain a holistic view that considers the many packaging design contributions to sustainable development. Consequently, it is not intended to be a reference book *per se*, but rather an inspirational guide to this complex and important topic.

In line with that ambition, the book aims to reach the minds of all professionals and companies that have, or do not have, packaging as a core competence or business. Packaging design is a cooperative team effort of people from multiple disciplines. Thus, there are many professionals from various disciplines, company functions and departments for that matter, who are involved in packaging design. This can include professionals from R&D, production, marketing, sales, finance, purchasing, logistics and regulatory. For a majority of these professionals, packaging is not their core competence. This book provides them with guidance so that they can navigate the packaging landscape. Yet, for highly experienced professionals in the realm of packaging design, the book provides great inspiration and valuable new ways of thinking.

Theory and practical applications are balanced by dividing this book into three integrated parts. In Part I, the basic tenets of packaging, sustainability and design are presented to make the book more managerial, integrative and "cutting edge". Views on sustainable development and packaging design are also subjects that you will become acquainted with in Part I. It "sets the scene" for what is to come: the packaging design compass for sustainable development. Part II is the focal point of the book. It describes the compass in detail, its directions and how to navigate with it. Part III exemplifies the compass directions in a wide range of illustrative cases that help readers to understand and gain insights into explorative, comparative and real-life cases. It aims to inspire and challenge the mindsets of those who apply the compass in packaging design related projects. The case material is integrative in nature and examines directions of the compass that are important for sustainable development. The cases are structured to inspire readers in the challenging task of packaging design thinking.

Packaging design for sustainable development is a field in its infancy, veiled behind preconceived myths and misconceptions. There is a tremendous amount of knowledge that needs to be generated and disseminated, and there is considerable interest from industries and academia to take in and apply this knowledge. *Managing Packaging Design for Sustainable Development – A Compass for Strategic Directions* is the only publication that takes a broad supply chain orientation and views the subject from a sustainable development perspective. While emphasizing the supply chain aspects of packaging, it integrates all three pillars of sustainable development as well as incorporating how to make strategic decisions in relation to packaging design. In addition to the compass, there are several important topics that are unique to this book or are approached in a new way. Examples are the complexity and challenges of packaging design, and the packaging logistics perspective as such. The multidisciplinary themes are interwoven throughout the chapters.

This is a must-have book for designers, engineers, logisticians, marketers, SCM professionals and other managers who seek guidance on sustainable solutions through packaging design. The nature of the book is pragmatic and applied in its approach to managing packaging design for sustainable development. It is also a valuable source of knowledge and practical experience for students, public officials, researchers, policymakers and many others who have a strong interest in packaging design and sustainable development. It fills the gap in the scarcity of books about the crucial role packaging design plays in sustainable development. It clearly takes a giant leap from thinking of "sustainable packaging" to thinking of "packaging design for sustainable development" by comprehending the whole rather than the separate parts.

As with packaging design processes, the outcome of this book has been a journey of iterations based on the authors' many years of experience. This has been intertwined with the practical cases and integrated with the existing but sparsely reported research in the field. The journey has taken several directions. It has consisted of real-life presence and off-road imaginary thoughts. At the end of this journey, we wished we'd had a compass. Yet without a compass, we still feel we have moved in the right direction by contributing to and inspiring packaging professionals and communities to strive for a more sustainable world.

Lund, on Leap Day, 29 February 2016
Daniel Hellström and Annika Olsson

Acknowledgements

There are many people and organizations that contributed to the research behind this book, and who supported us in different ways during its actual writing.

The Bo Rydin Foundation has been the first and foremost funder from the start. Its donation to Lund University in 1994 founded packaging logistics as an educational and research subject area. The research group has since grown and established itself and is internationally recognized. This book is based on this original packaging logistics research. Without the initial donation and ongoing funding from the Foundation, neither the area of packaging logistics nor this particular book would have come into being.

The idea for the book came up during a research project funded collaboratively by two Swedish funding bodies, "Formas" and "Handelns utvecklingsråd", under their programme "Sustainable Retail". We were specifically funded for a project called "Packaging Design for Sustainable Development of Retail". As a result, we developed the first prototype of the compass and investigated, gathered and authored all the illustrative cases in the book with our colleague, Professor Fredrik Nilsson.

We are grateful to all the people and companies for the time spent sharing their insights about the cases and for providing us with the case material. We have specifically acknowledged your support in conjunction with each case.

Throughout the overall development process of the book, a number of people were very helpful. Special thanks go to Erik Andersson, our supportive colleague whose photographic skills have illustrated our research. Erik's patience with us, our alterations and our short deadlines has been amazing and much appreciated. We are very grateful to Eileen Deaner. Being native Swedes, we do our best to write in English, but with a never-ending support, she turned the text into something understandable for an international audience. Thank you Eileen for working days and nights to improve our writing and for never giving up on us! We are also grateful to Catrin Jakobsson for the well thought through illustrations. Catrin was able to understand what we wanted to illustrate and make it better than we had ever expected.

This book has been a mix of pain and pleasure for us to write and assemble. Without the support of our families, it would have been impossible. The first author wishes to thank his soul mate, Josefine Broman, who has been a constant source of inspiration and support in maintaining a balanced life. Special thanks goes to the first author's son, Hjalmar, and daughter, Lisa, for giving up hours of time with their father so that he could work. The second author would like to thank her family who has lived the last year with her in parallel with this book project. Her guiding stars of life are Torben, Elin and Anton. Thank you for the inspiration, patience and support you have provided, one of the many reasons why I love you all.

Finally, to all our academic colleagues. We are indeed grateful for the daily discussions about research studies and projects, including this book. Co-creation often occurs in these discussions, a co-creation that advances our knowledge in the field we are dedicated to. See you at the coffee machine!

Part I

Fundamentals of Packaging Design

Packaging is something that we interact with on a daily basis. Most of the time we do not even notice it, since packaging is fully integrated into our lives and personal use, as well as with the product inside. Can you imagine what the world would be like without packaging? Packaging ensures that the products of the world reach the consumers of the world. Some people may argue that because packaging is not part of the product, it is not needed and should be restricted or even banned. When we eat, we are safeguarded because our food has been protected by packaging. When we are sick, our pharmaceuticals are safe, efficient and not counterfeit, due to effective packaging. There would be no need for packaging if the products themselves were resistant to everything in all types of surroundings, if they did not have to be moved, and if they were not time dependent. But as we all know, this is not the case nor will it be in the future. This is why packaging is a prerequisite for safe production, distribution and consumption.

In many parts of the world packaging is an intrinsic part of businesses, industries, institutions and authorities. In businesses, packaging plays an important role in the renewal and extension of product life cycles and is recognized for its positive effects on productivity, its financial impact and its value creation. Packaging is a global business with an annual turnover of close to €500 billion that is growing in line with the global economy. For governments, packaging does not only affect the national economy but also its legislation. For society as a whole, packaging is a vital element in enabling population growth, fostering new and changing habits and life styles, creating employment and trade and most importantly, contributing to the availability of products around the world.

Part I of this book – *Fundamentals of Packaging Design* – is made up of three chapters. The first, "Introduction to packaging", is where the functions, legislation, regulations and terminology of packaging are explained to introduce you to the world of packaging systems. The chapter ends by describing the multidisciplinary nature of packaging and the role of packaging logistics. Chapter 2 is about "Sustainability development and packaging". Here we present the definition of sustainable development, the historical role of packaging, and common misunderstandings about packaging. We also elaborate on how packaging can effect and encourage sustainable development. In Chapter 3, "Designing packaging", packaging design is examined from various points of view to explore the "brilliance" and complexity of its numerous aspects and facets. The management, practices and tools of the packaging design process are also presented. The numerous requirements and needs of packaging are described, followed by the design challenges in dealing with this complexity.

1 Introduction to packaging

Packaging is the science, art and technology of protecting and adding value to products. In order to fulfil these tasks, it is necessary to integrate the processes of designing, evaluating and producing packages, which also involves the elements of materials, machinery and people. People have a variety of views on packaging. One of the more limited views is reflected in the question: What packaging material is better than another? In reality, material is only one element of packaging, one which is highly dependent on the product that is about to be packed. This limited view needs to be supplemented by others in order to take in all the different perspectives of packaging and the functions it has throughout its life cycles. To clarify the meaning of packaging, a broad and well-established packaging definition is needed. The definition we use in this book is based on Paine's (1981) well established version and the EU's definition (94/62/EC). It is expressed in three statements:

1) Packaging is a coordinated system made up of any materials of any nature, to be used for preparing goods for containment, protection, transport, handling, distribution, delivery and presentation.
2) Packaging is the means of ensuring safe delivery from the producer to the ultimate consumer in sound and safe conditions.
3) Packaging is a techno-economic function aimed at making delivery efficient while maximizing effectiveness.

The package itself is defined as the physical artefact that performs the many functions required from different stakeholders and from the product. This is our jumping off point for further elaboration on the different functions of packaging.

1.1 Multiple functions of packaging

The principal functions that packaging is able to perform are manifold. Several authors and researchers in the packaging field have described and defined them in various ways. Paine (1981), Robertson (1990) and Livingstone and Sparks (1994) emphasize seven fundamental functions of packaging for the product

Managing Packaging Design for Sustainable Development: A Compass for Strategic Directions,
First Edition. Daniel Hellström and Annika Olsson.

to be: protection, containment, preservation, apportionment, unitization, convenience and communication of the product. Lockamy III (1995) lists the same functions, but excludes preservation, which mainly relates to food and other perishable products. In Lockamy III's assessment of strategic packaging decisions, the six main functions of packaging are: containment, protection, apportionment, unitization, convenience and communication. These six fundamental functions are the ones that most researchers acknowledge and use, even though some of the functions have been developed and expanded. For example, the protection function can be divided into physical and barrier protection. Others researchers integrate functions by merging the above-mentioned six into broader categories. Lindh et al. (2016) propose three main functions: protect, facilitate handling and communication. Another way of categorizing packaging is to use process-related aspects such as security, marketing and information transmission as specific functions. One can claim, though, that security can be sorted under the protection function, as well as under communication; marketing and information transmission can also be sorted under communication (Lindh et al., 2016).

We could take any of the above-mentioned set of functions as our starting point, but have chosen Lockamy III's (1995) six main functions because they are the most commonly used and referred to. We have also added information as a function of its own.

1.1.1 Containment

The purpose of containment is to hold the content and keep it or the surroundings secure. The second part of this definition is similar to protection, but more clearly signals the activity of collecting things into an assembled unit. Many products need containment because of their nature, the classic example being liquids. Since products come in all shapes and sizes and react in different ways to their surroundings, some kind of containment is necessary. Imagine the process of getting pasta or rice to your dinner table without packaging. Containment highlights the need for the existence of packages in making products available to consumers.

1.1.2 Protection

The protection function of packaging involves safeguarding the contents of the package from external sources and vice versa. Damage can arise from physical, chemical, microbiological and climatic sources. Packaging provides physical protection against many different static and dynamic forces, such as vibration, compression and mechanical shock. It also protects from climatic conditions and hazards, such as temperature and humidity. From a chemical and biological point of view, it protects the product from microbiological or chemical deterioration, which is also a preservation function. Preservation means retaining the quality of the content by stopping or inhibiting chemical and biological changes. It can be regarded as part of the protection function because it is usually managed by choosing a proper packaging material. Preventing damage from external sources is often considered the main reason for having packaging.

But the package also functions as protection of humans and the environment from the internal product. An example of this is in the transport of hazardous materials.

1.1.3 Apportionment

The apportionment function enables a given amount of content to control and support appropriate usage. Apportionment in packaging facilitates the output from today's large-scale industrial production by dividing products into manageable portions and sizes. This provides retail outlets and consumers with the desired amount and proper dimensions of the product for different users in different situations. Apportionment also helps users to manage inventory and to reduce food waste by using appropriate portion sizes. Apportionment is similar to and meets the same underlying needs as the next function, unitization.

1.1.4 Unitization

Unitization involves the consolidation or reconciliation of units. Most often small units are grouped into bigger ones to improve efficiency. However, large units are regularly divided into smaller units to be assembled later and elsewhere. Like apportionment, it helps to make the handling of packaging suitable for different stakeholders in different situations and at different locations. The primary function of the famous shipping container, sometimes called "the box that made the world smaller and the world economy bigger", is unitization. Unitization is sometimes used as a synonym for agglomeration.

1.1.5 Convenience

The primary purpose of the convenience function is to make it easy and convenient to use the packaging and its contents. Convenience relates to unitization and apportionment. The main task of all three is to facilitate handling and to package the product in appropriate sizes and amounts for its use at different stages in its life cycle. This can be done in practically endless ways. Aspects of the convenience function in packaging throughout all stages in production and distribution to the final consumption and recovery include:

- the ability to consume products at any time and any place;
- the perception of the packages as being easy to open, carry and empty;
- providing accurate and safe dispensing;
- ease of disposal.

1.1.6 Information

The package is the interface between the product and the logistics, and between the product and the consumer or other users. This means that the role of the package as the information carrier is essential. Information on packages is often taken for granted, but sometimes underestimated and forgotten. We need to keep in mind that information constitutes a fundamental function of packaging.

The two major roles of packaging information are to help users identify the content, and to provide them with instructions on how to use it. Barcode technology is a ubiquitous element of modern civilization and an integrated part of packaging information. Other technologies applied to packaging indicate tampering. These include authentication seals, security printing and other anti-theft devices on packaging that tell you the package and content are not counterfeit or stolen; they also serve as a measure of loss prevention. The interactive information achieved through this technology development is part of the communication function in packaging, which is a type of two-way information between the producer and the user.

1.1.7 Communication

The package is a medium for communication between the brand owner and the consumer. This kind of communication is regarded as marketing. Packaging is sometimes called "the silent salesman", especially when it comes to groceries. Packaging is often the first and most regular contact consumers have with a product, attracting the eye and whetting the appetite. More than just giving a face to the brand, packaging is a powerful sales and commercial tool. It influences market position and consumer behaviour by triggering purchase and creating identity and loyalty. Marketing communication in the form of physical and graphical design is often applied to the package in order to bring products to life in accessible and engaging ways from the way they look, feel and function, to how the content is perceived.

But just as packaging can be seen from different points of view than those of the product and packaging developers, so can its functions. For example, packaging directly and indirectly impacts different organizational business functions such as:

- logistics (handle, transport, store, distribute, inform);
- marketing (sell, differentiate, promote, provide value, inform);
- production/manufacturing (produce, make, assemble, fill);
- information systems (perform, inform);
- environment (reduce, reuse, recover, dispose, inform).

It can thus be argued from an organizational business perspective that these processes are the main functions of packaging.

1.2   Packaging legislation and regulations

The legal ramifications of the initiatives taken by governments and authorities can impact the way actors relate to the packaging functions described in section 1.1. The legislation that affects packaging comes in many forms and subjects because there is no separate branch of "packaging law". Examples of the legislation concerned with packaging cover the sale of goods, transport, environmental issues, food and drugs, food safety and waste management. The legislation and regulations are constantly under revision and updated frequently.

Corner and Paine (2002) provide an excellent overall categorization of the areas in which many countries have packaging legislation. The most important areas fall under the following four categories:

1) *Administrative needs:* For example, regulation for food packaging, pharmaceutical and medical packaging and dangerous substances.
2) *Requirements to protect the public:* For example, child-resistant packaging, tamper-evident packaging, fraud, and weight and measurement directives.
3) *Protection of packaging designs:* For example, copyright laws, intellectual properties, trademarks and patents laws.
4) *Environmental protection:* For example, packaging and packaging waste directives and producer responsibility

The intention here is not to give a complete list or description of laws and regulations, but to provide examples of significant legislation and regulations found in the categories that apply to packaging. Hence, an overall description is presented.

1.2.1 Administrative legislation and regulations

Two areas with extensive legislation on the interactions between the packed product and the materials in which it is packed are food and packaging compatibility, and food and drug material contact. Currently, one major material is plastic and the legislation is mostly concerned with the contact plastic has with food; but in principle other materials are also applicable. The legislative concerns are primarily if substances migrate from the packaging into the foodstuff or drugs, and are either harmful for the consumer or have an adverse effect on the contents' deterioration properties – such as taste and aroma. However, any kind of inert migration and contamination is undesirable. The US Food and Drug Administration and the European Union are authorities that regulate this important legislation.

1.2.2 Legislation and regulations for protecting the public

There are laws that regulate labelling and consumer information to protect the public. Consumer information must not mislead the consumer in regard to the contents' nature, properties, composition, quantity, origin and durability. In the European Union, the Energy Labelling Directive (2010/30/EU) and the Ecolabel Regulation (EC, No. 66/2010) concern packaging, or at least the content inside the package. Another important law is that of child-resistant closures for packages containing dangerous substances sold to the general public. For some products, there are regulations about tamper-resistant packaging, which means that it must have an indicator or barrier to entry that is distinctively designed, or must employ an identifying characteristic (a pattern, name, registered trademark, logo or picture).

1.2.3 Legislation and regulations for protecting designs

Some copyright, intellectual properties, trademarks and patent laws fall into the category of protecting packaging designs and technologies. Trademarks may comprise words, letters, numerals, names, designs, or the shape of goods or their packaging. To be entitled to packaging trademark protection, a trademark has to be distinctive enough so customers can identify the packaging with a company product or service and not a competitor's. Patents are registered in specific countries and are valid for a set period of time before they are released for use by others.

1.2.4 Legislation and regulations for protecting the environment

There are two major established environmental policies related to packaging in Europe. Their underlying principles are that preventive action should be taken, environmental damage should be managed at the source, and the polluter should pay. The EU Packaging and Packaging Waste Directive (EU, 2015/720) seeks to provide environmental protection and to ensure the functioning of the internal EU market. According to the Directive and its appendix, packaging must meet certain essential requirements and member states must ensure that packaging placed on the market complies with these requirements:

- to limit the weight and volume of packaging to a minimum in order to meet the required level of safety, hygiene and acceptability for consumers;
- to reduce the content of hazardous substances and materials in the packaging material and its components; and
- to design reusable or recoverable packaging.

Another major environmental policy related to packaging is the Extended Producer Responsibility (EPR). This is defined and described by the OECD as:

> ...an environmental policy approach in which a producer's responsibility for a product is extended to the post-consumer stage of a product's life cycle. An EPR policy is characterized by: 1. the shifting of responsibility (physically and/or economically; fully or partially) upstream toward the producer and away from municipalities; and 2. the provision of incentives to producers to take into account environmental considerations when designing their products. While other policy instruments tend to target a single point in the chain, EPR seeks to integrate signals related to the environmental characteristics of products and production processes throughout the product chain.

With the Extended Producer Responsibility, the producer pays in advance for the pollution the future owner will generate. In this way, consumers are relieved of the responsibility for the disposal of packaging.

In contrast to Europe, the US Federal Government has not published any national environmental packaging regulations or introduced extended producer responsibility. In the 1990s, the US President's Council on Sustainable Development introduced the term *extended producer responsibility* to mean that all participants in the product life cycle (governments, companies with economic interest in the product, consumers, and those handling waste) share the responsibility for the environmental effects of products. This difference in definition and use of the extended producer responsibility concept from the Europe version has yet to find its way to any legislation.

1.3   Packaging terminology

In order to gain a coherent understanding of packaging as defined by Paine, it is best classified as primary, secondary or tertiary (Paine, 1981). This classification should be used when packaging is regarded as a system and illustrates the levels of hierarchy in the packaging system (Figure 1.1).

However, several other terms than the above-mentioned levels are used by practitioners in different industries when discussing different types and levels of packaging. This incoherent use does not make communication easier and often causes miscommunication and difficulties in understanding packaging. Table 1.1 summarizes some of these terms and shows that a term often describes both its primary function and use.

Other terminology used in the description of the package relates to the type of product it contains: food packaging, pharmaceutical packaging, etc. In short, there are numerous viable terms used in packaging. This calls for a more homogeneous and systematic terminology for classifying packaging: primary, secondary and tertiary packaging.

Theses broad classifications of packaging components are based on viewing packaging as a system, and are somewhat arbitrary. For example, is the shrink wrap of a six pack of soda cans primary or secondary packaging? There is no clear-cut definition or borderline between the three different levels of the hierarchy,

Figure 1.1 The levels of packaging: primary, secondary and tertiary packaging.

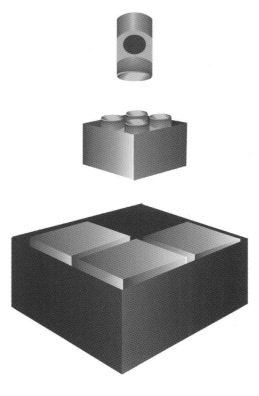

Table 1.1 Examples of terms used for packaging.

Packaging type	Description
Consumer packaging	The packaging that the consumer usually takes home.
Sales packaging	The unit for sale.
Group packaging	Packaging that is created to facilitate protection, display, handling, and/or transportation of a number of primary packages.
Retail packaging	Same as group packaging with an emphasis on the design to fit in retail.
Display packaging	Same as group packaging, often with an emphasis on the display features.
Transport packaging Industrial packaging Distribution packaging Bulk packaging	Packaging that facilitates handling, transport and storage of a number of primary packages in order to provide efficient production and distribution, and to prevent physical handling and damage during transportation.
Used packaging	Packaging/packaging material remaining after the removal of the product it contained.

and thus no exact right or wrong. In many cases it is all up to what you define as being included in each component of the packaging system.

1.4 Packaging as a system

In trying to understand and explain the world of packaging, we consider it as a system represented by the product and different levels of packaging – primary, secondary and tertiary. These levels are interrelated and affect each other (Figure 1.2), which means that the levels and the interactions cannot be regarded or assessed on their own. So if any changes are made in the primary packaging, it may not only affect the product and secondary packaging, but the tertiary packaging as well. This line of thinking is found in systems theory, which plays a central role in a wide range of scientific fields. The concept of a system is described by Checkland (1999: 3):

> The central concept "system" embodies the idea of a set of elements connected together which form a whole, this showing properties which are properties of the whole, rather than properties of its component parts.

Systems theory stresses holistic thinking in order to prevent reductionism, and is based on the assumption that the whole (system) is not necessarily equal to the sum of its parts (Churchman, 1968; Von Bertalanffy, 1969). The interactions among the parts forming the system can make the sum of the system greater or lesser than the sum of the parts; a change in one part can have a negative or

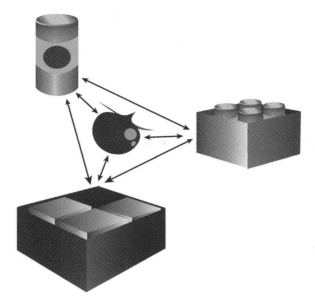

Figure 1.2 The packaging system and its interactions among the packaging levels and the product.

positive impact on other parts of the system. So if you want to gain insight into the performance of a system, you need to understand the interactions among its parts. Do this rather than reduce the system to separate parts to be analysed on their own. The core of systems theory is how different parts interact with other parts of the system and what this means for the entire system.

By applying systems theory to packaging, the interactions between the different packaging components are highlighted. It explains and helps us understand the interdependence among the components. The performance of the packaging system is not only affected by the performance of each individual packaging component, but also by the interactions among them. Considering these interactions is critical to the overall performance of the packaging system.

In line with this reasoning, you need to remember that the product is an interactive component in the packaging system (Esse, 1989; Griffin et al., 1985). It obviously interacts physically with the primary packaging, but it also interacts in several other ways with the other two packaging components. A paper-based package of crackers or a bag of potato chips, for example, require a rigorous secondary package to protect the content from mechanical damage, because the primary packages are too weak for that purpose.

Being able to distribute the functions so that they are not equally performed across all packaging components is a typical example of how the components are interrelated and must be seen as a "whole". As in the example of the crackers and chips, the primary packaging is able to preserve the product while the secondary packaging provides the mechanical protection. In this sense, certain choices/decisions made at one level are likely to affect all the other components.

1.5   Packaging goes beyond a single discipline

Packaging is a multidisciplinary academic field. This means that experts from different disciplines meet and work together based on their perspectives and expertise in packaging related issues. They retain their discipline's methods and assumptions and contribute with in-depth knowledge linked to that of the other disciplines in the multidisciplinary relationship. In academia, the field of packaging is becoming increasingly attractive to students and scientists in the engineering sciences (including but not limited to industrial management, mechanical engineering, production, chemistry, and product development), food sciences, pharmaceutical sciences, social sciences, materials science, design sciences, environmental science and business-related disciplines (including business administration, marketing and management sciences).

In contrast to academicians, packaging professionals usually interact multidisciplinarily with people in different organizational functions such as R&D, production, marketing, sales, finance, purchasing, logistics, regulatory and more. These interactions occur internally in an organization and externally with consultants, suppliers, customers and end users. A majority of the professionals working with packaging or making packaging decisions, however, do not see themselves as being packaging professionals. This is because they have an affiliation and sense of belonging to other organizational departments such as product development or marketing. These professionals have different views, sometimes even conflicting opinions, on the packaging system and the functions it should carry out, depending on what department they represent. For example, primary packaging has always been in the realm of the marketing department, while secondary and tertiary packaging have been in the realms of the production, manufacturing or transport departments. Outside of their organizational affiliations, they rarely understand the impact their packaging decisions have on the whole packaging system or on suppliers, customers and end users along the entire supply chain. A package might look great, perform excellently in the production line, but the impact it has on a bigger scale, such as that of sustainable development, is seldom reflected on and yet it is one of the major challenges facing packaging design. Clearly, packaging needs to function properly in all stages from production to consumption and recycling in order to contribute to the greater whole. If the packaging is difficult to produce, fill, close and collect in the packaging line, it does not matter that it has an effective structural design from a protective point of view or good graphics that promote and sell the product.

There is no question that packaging has many facets. Apart from its material, physical properties and structural design, it is in itself a valuable tool, especially in logistics and marketing. It is an essential link between the producer and the consumer, where it contributes to the positioning and presentation of the product; and on many occasions, the use of the product after purchase. These many facets need to be represented by multiple roles in businesses, since packaging needs to meet the requirements placed on it from logistics, marketing, production, product development, and from the environment. This calls for an interdisciplinary approach to the study of packaging systems, where disciplines are interrelated both in content and methods. Five aspects of a single but interdisciplinary process are developing a product, designing a packaging system, producing, distributing and marketing the product. This is the foundation on which the interdisciplinary field of packaging logistics has been established.

1.6   Going multidisciplinary – packaging logistics

Packaging logistics primarily brings together different packaging disciplines with logistics and supply chain management disciplines to complement and support one another. This multidisciplinary field of study crosses the traditional boundaries between academic disciplines and schools of thought as new needs and professions have emerged. Packaging logistics focuses on the synergies achieved by integrating the systems of these disciplines, with the aim of adding value to the combined, overall, system. The core of packaging logistics covers the design of a product and its packaging system, throughout the whole supply chain from raw product, via various actors, to the end user, and on to recycling and recovery. It would be difficult to develop a sustainable logistics system without packaging that supports it or, vice versa, to create a sustainable packaging system without the support of the logistics and supply chain management.

Logistics is an application-oriented management discipline of the "flow of things". According to the Council of Supply Chain Management Profession (CSCMP), formerly the Council of Logistics Management, logistics management is defined as:

> …that part of supply chain management that plans, implements, and controls the efficient, effective forward and reverses flow and storage of goods, services and related information between the point of origin and the point of consumption in order to meet customers' requirements.

In a simple description, logistics aims to achieve "The 7 Rights of Logistics": having the *right* item in the *right* quantity at the *right* time at the *right* place for the *right* price in the *right* condition to the *right* customer. Logistics management is one part of supply chain management, and that has a larger scope. The CSCMP definition is as follows:

> Supply chain management encompasses the planning and management of all activities involved in sourcing and procurement, conversion, and all logistics management activities. Importantly, it also includes coordination and collaboration with channel partners, which can be suppliers, intermediaries, third party service providers, and customers. In essence, supply chain management integrates supply and demand management within and across companies.

Even though packaging is included in logistics and supply chain management to varying degrees, it is recognized by scholars as having a significant impact on logistics costs and performance (Ebeling, 1990; Twede, 1992). Bowersox and Closs (1996) concluded that packaging affects the performance of every logistical activity throughout the supply chain, either directly (material handling and transportation) or indirectly (as information carrier, product protection, etc.). Nevertheless, packaging is often regarded in logistics as an unavoidable non-value-added cost containing little to no strategic value (Lockamy III, 1995). This has resulted in the packaging-dependent costs in the logistics system being frequently overlooked by packaging and logistics professionals (McGinnis and Hollon, 1978; Twede, 1992).

Gattorna (1990) presents the role of packaging in logistics as "this long-neglected but fundamental part of our activities" and states that packaging is a source of profit and, in fact, also has an impact on the environment. In the logistics and supply chain management literature, these have essentially become the two main packaging themes: source of profit and environmental impact.

References

Bowersox D.J. and Closs D.J. (1996), *Logistical Management – the Integrated Supply Chain Process*, International Edition. McGraw-Hill, New York.

Checkland P. (1999), *System Thinking, System Practice – includes a 30-year Retrospective*. John Wiley & Sons, Chichester, UK.

Churchman C.W. (1968), *The Systems Approach*. Dell Publishing, New York.

Corner E. and Paine F. (2002), *Market motivators – the Special Worlds of Packaging and Marketing*. CIM Publishing, Berkshire, UK.

CSCMP, Council of Supply Chain Management: https://cscmp.org/

Ebeling C.W. (1990), *Integrated Packaging Systems for Transportation and Distribution*. Marcel Dekker, New York.

The Ecolabel Regulation (EC No. 66/2010): http://eur-lex.europa.eu/LexUriServ/LexUriServ.do?uri=OJ:L:2010:027:0001:0019:en:PDF

Esse R.L. (1989), Package development, manufacturing, and distribution strategy consideration. In: *Packaging Strategy Meeting the Challenge of Changing Times*. A.W. Harckham (ed.), Technomic Publishing AG Lancaster, Pennsylvania, pp. 107–116.

The Energy Labelling Directive (2010/30/EU): https://ec.europa.eu/energy/en/topics/energy-efficiency/energy-efficient-products

The EU Packaging and Packaging Waste Directive (EU, 2015/720): http://ec.europa.eu/environment/waste/packaging/index_en.htm

European Parliament and Council Directive (94/62/EC): http://ec.europa.eu/environment/waste/packaging/index_en.htm

Gattorna J.L. (1990), *Foreword in special issue of International Journal of Physical Distribution and Logistics Management*, 20(8), 1–4.

Griffin R.C. Jr., Sacharow S. and Brody A.L. (1985), *Principles of Packaging Development*, 2nd Edition. Van Nostrand Reinhold Company, New York.

Lindh H., Olsson A. and Williams H. (2016), Consumer perceptions of food packaging: Contributing to counteracting environmentally sustainable development. *Packaging Technology and Science*, 29, 3–23.

Livingstone S. and Sparks L. (1994), The new German packaging laws: Effects on firms exporting to Germany. *International Journal of Physical Distribution and Logistics Management*, 24(7), 15–25.

Lockamy III A. (1995), A conceptual framework for assessing strategic packaging decisions. *The International Journal of Logistics Management*, 6(1), 51–60.

McGinnis M.A. and Hollon C.H. (1978), Packaging, organization, objectives and interactions. *Journal of Business Logistics*, 1(1), 45–62.

OECD, Extended Producer Responsibility (EPR): http://www.oecd.org/env/tools-evaluation/extendedproducerresponsibility.htm

Paine F.A. (1981), *Fundamentals of Packaging*. Leicester: Brookside Press.

Robertson G.L. (1990), Good and bad packaging: Who decides? *International Journal of Physical Distribution & Logistics Management*, 20(8), 37–41.

Twede D. (1992), The process of logistical packaging innovation. *Journal of Business Logistics*, 40(4), 85–88.

Von Bertalanffy L. (1969), *General Systems Theory Foundations, Development, Applications*. George Braziller, New York.

2 Sustainable development and packaging

Some of the most intriguing tasks for sustainable development in the world today are to safeguard the limited natural resources on our planet and to stabilize the concentration of greenhouse gases in order to mitigate climate change. This needs to be achieved in parallel with careful considerations for eliminating poverty and ensuring prosperity for all. Large portions of the resource use and the greenhouse gas emissions have their origins in the production of goods that we consume and use in our daily lives. And since the majority of goods are packed, packaging certainly affects both reource use and greenhouse emissions in different ways. At the same time as reources are becoming scarce, parts of the population have limited access to the reources necessary for their survival, such as food and water. The current situation places some intriguing challenges on our society, some of which are directly or indirectly related to packaging. One major challenge, where packaging can make an indirect contribution, is the availability and safe distribution of scarce resources such as food, drinking water and pharmaceuticals for a growing population. Another challenge directly related to packaging is that more than 6.5 million tons of plastics are dumped each year in the oceans and a lot of packaging waste ends up in landfills. Most of these packages are made of finite resources. One way to reduce greenhouse gases is to increase the use of renewable raw materials or recycled materials, in addition to carefully using and protecting the limited natural resourses.

A third challenge indirectly related to packaging is transportation that results in emissions, congestion and noise. Packaging indirectly affects volume and weight which can reduce transportation dramatically and in turn contribute to reducing greenhouse gases. This is certainly needed because European emissions have to be reduced by 70–90% in order to reach the zero-level sustainability goals after year 2050 (MBV, 2007).

2.1 Sustainable development goals

Many organizations and businesses of today have acknowledged the problems we face with climate change and over usage of the world resources. They have realized the need to work for a sustainable society. Numerous business visions

Managing Packaging Design for Sustainable Development: A Compass for Strategic Directions,
First Edition. Daniel Hellström and Annika Olsson.
© 2017 John Wiley & Sons, Ltd. Published 2017 by John Wiley & Sons, Ltd.

also include sustainable strategies and there are many organizations that actually do work with sustainable considerations. When speaking about sustainable societies, it is interesting to reflect on the word *sustainable* and what it means. In its adjective form, *sustainability* indicates a condition that is about maintaining a certain rate or level, meaning maintaining a present state.

But rather than maintaining the present state, we think that in order to acknowledge sustainability, continuous work is needed for a better future. That is why the words *sustainable development* are more suitable. They indicate that this is something in progress; a process that is continuously in motion and hopefully moving in a positive direction. It is also a process in which you never can relax! If considered as a process, sustainable development can be affected by different initiatives taken by different actors in a system; it needs to be assessed over time and from the point of view of the entire system. One actor can therefore not only care about their own actions, without appreciating what is happening with other actors in the system. For example, a retailer or brand owner needs to consider if their suppliers (i.e. the product manufacturers) use or carry dangerous chemicals or if they are using child labour. They also need to consider how their products are transported to the point of sale along with other aspects included in guidelines for corporate social responsibility (Carroll and Shabana, 2010; McWilliams et al., 2006).

The concept of sustainable development was first defined by The World Commission on Environment and Development in 1987 and is one that is well-known and so often used:

> Sustainable development is development that meets the needs of the present without compromising the ability of future generations to meet their own needs.

Following the 1987 World Commission, there have been several global meetings about and many initiatives have been taken for sustainable development. In 1992, the international community gathered at the Rio Earth Summit to discuss how to operationalize sustainable development. It was here that top leaders adopted Agenda 21. This agenda contained specific action plans for realizing sustainable development at national, regional and international levels. Built on the progress made and the lessons learned from Agenda 21, the Johannesburg Plan of Implementation was then adopted at the 2002 World Summit on Sustainable Development. This implementation plan had concrete steps and quantifiable and time-bound targets and goals. It was also here the promotion of sustainable patterns of consumption and production was identified as one of three overaching objectives of sustainable development (Fitzpatrick et al., 2012).

In addition to these initiatives, the member states of the UN created the *Millennium Development Goals* for reducing poverty and hunger and getting the world to work towards a more sustainable resource utilization. As a follow-up, the member states agreed on seventeen sustainable development goals in September 2015 that are better and more precisely defined than the previous eight. These global goals are set to end poverty, to protect the planet, and to ensure prosperity for all as part of a new sustainable development agenda.

Figure 2.1 The UN sustainable development goals for 2015.

Figure 2.1 summarizes these seventeen sustainable development goals.

The initiatives brought up at all the summits take an overall perspective on sustainable development. This perspective is both relevant and necessary, since sustainable development has to be viewed and implemented from a holistic systems perspective. And it is a global matter! But, for people and businesses to be able to contribute to sustainable development, the goals need to be taken down to operational levels. This has to be done in order for businesses and people to develop and deliver the products and services required to meet the goals of social responsibility, environmental conservation and economic profitability (Elkington, 1997).

2.2   Three pillars of sustainable development

Since the concept of sustainable development was first defined in 1987, sustainable development has emerged as the guiding principle for long-term global development. It is based on the three pillars of sustainability or the triple bottom line (3BL) of sustainability as it is also called (Elkington, 1997). The three pillars aim to achieve social equity, environmental protection and economic prosperity in a balanced manner (UN website, 2015). The guiding principles for sustainable development thus require social, environmental and economic considerations in all developments in society.

Based on the integrating nature of the three pillars of sustainable development – social responsibility, environmental conservation and economic profitability – the terms "people, planet and profit" are often used. The social responsibility pillar equals ***people***, the environmental conservation equals the ***planet*** and the economic profitability equals ***profit*** (Figure 2.2).

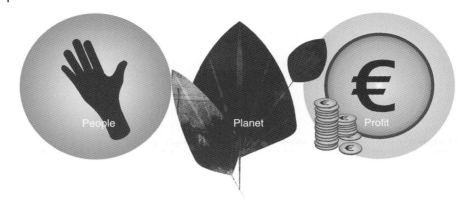

Figure 2.2 People, planet and profit for sustainable development.

2.2.1 People

The people pillar is about working for social equity which includes the aspects of healthy, safe and democatric societies. Businesses need to look into these sustainable development goals and identify what they mean for their organizations. The goals related to the people pillar should be viewed from different business process perspectives such as production, distribution or innovation, but also from the working environment perspective. On the part of people, this comes down to the social responsibility that organizations need to take, not only for themselves but for their entire business system of suppliers and customers and ultimately the end consumer. This means incorporating social responsibility into all business processes. In addition, organizations need to ensure that the people in the system maintain social equity and welfare for workers in the production, distribution and handling of these products and services.

2.2.2 Planet

Environmental protection and conservation means to carefully use the resources on the planet, without compromising for future generations. Responsible production means to make use of as many renewable resources as possible and think about reusing and recycling resources for circular economies. The planet pillar is also about preventing and reducing environmental pollutions, such as greenhouse gases. That means to promote susitanable production and consumption of products and services. Even for this goal, businesses need to look into their own business processes along with those of their suppliers and customers, in order to be able to take an overall responsibility for careful resource use and sustainable processes.

2.2.3 Profit

Economic prosperity refers to organizations in a system needing to survive and make a profit for their shareholders in order to carry out decent work and achieve economic growth, while still promoting sustainble production and consumption.

Engaging in sustainable development means that organizations have to do so without compromising the long-term survival of the business. One step to meet these sustainability goals is for industries to work with innovations and for municipalities and governments to support these efforts. Organizations also need to contribute to responsible production and consumption while maintaining their long-term survival. The profit pillar, or economic prosperity of sustainable development, also means contributing to ending poverty. This means providing affordable products and services, but it can also mean developing completely new business models and to provide high-quality employment.

The three pillars of sustainable development are elaborated individually above. Yet all three pillars need to be considered in the same assessment, meaning that the products that are developed, produced and distributed need to be economically effective, and result in profits for the organizations involved. At the same time, the same product needs to be produced, distributed and sold in an amount completely used for its purpose in a careful manner, as part of environmental effectiveness. In addition, it should reach the people in need or desire, in order to improve their lives and the lives of the interrelated stakeholders for societal effectiveness. Taking all three pillars into consideration should be done with the overall purpose of "meeting the needs of the present without compromising the ability of future generations to meet their own needs."

2.3 Looking back at the role of packaging

As early as the second half of the 1800s, self-supporting production of goods in the US turned from being home-grown and handcrafted to becoming industrialized (Twede, 2012). In Europe, this turnover had already started and packaging was being developed, but not on a large scale. With the industrial transition, production moved in place and time away from the final user, who evolved from being a self-supporting producer to a consumer who had to purchase goods that were distributed from the production location to the points for purchase. With this turnover, new needs arose for packaging to protect and help distribute products to the places of consumption. The role of the packaging to protect the products and to distribute them contributed to more people at more locations being able to benefit from products. Products became available to more people in resource efficient ways, which in turn led to better social and economic conditions, and the package was regarded as a necessity and contributor to the welfare of people. With the message: "Mother Earth makes some outstanding nuts, seeds and fruits. She just needs a little help with the distribution," the package in Figure 2.3 signals the necessity of packaging.

From that point in time, mass production and marketing of products started and self-service stores developed in the early 1900s. The first supermarket was introduced in the US in 1920. The way of buying groceries and other fast-moving consumer goods changed and opportunities arose for brand owners to market their products to consumers via the package. Instead of buying in returnable packages at several specialty stores with a dedicated sales person, all kinds of products were now sold in one self-service store without a specific sales person.

Figure 2.3 Package that illustrates the necessary role of distribution.

This meant that the package developed from just being a product protector to becoming "a silent salesman" and the carrier of the brand name; in other words, the natural interface between the product producer and the consumer (Beckeman and Olsson, 2005; Olsson and Larsson, 2009). In the role as the silent salesman, the first packages were printed with information about the content to help the consumer choose the right product in the purchasing situation. But as time passed, the brand owners realized the strategic potential of the marketing role of the package, leading to a rapid development of new graphics containing both factual information as well as pure marketing graphics to attract the attention of the consumer. This led to the development of new products in the same category and their packaging differentiation in terms of shapes, colours, print, etc. It also meant an exponential increase in products of the same kind, an increase that in those days neither considered sustainable production nor sustainable consumption.

The development of retail outlets for different consumer products has accordingly also increased heavily since the early 1900s and new developments are still to be found everywhere. This means that products and their packaging are travelling longer distances and to more dispersed locations. Retail has also developed from being just one physical channel – retail stores where you purchase your products and bring them home – to a multichannel structure with e-commerce as a parallel channel for products purchased over the Internet arriving at your home or at a collection point based on an individualized home delivery system.

The change of products in our world due to more differentiation, more complex retail structures and globalization have not only changed the role of the primary packaging but that of the secondary packaging as well. This has gone from distribution in returnable packaging that dominated retail in the past, to the different kinds of one-way secondary packaging solutions that are most commonly used today for distributing goods to physical retail stores. One reason for this development is that the supply of products has become a global matter, which makes one-way packaging more efficient for long-distance distribution and also better because of the lack of standardization for returnable packaging systems over country and continent borders. With the onset of e-commerce, the one-way system for packaging in home delivery has further increased the market for distribution packaging. However, in the e-commerce channel you can see new developments of returnable packaging in an effort towards more sustainable solutions. In this case, it is predominantly in returnable primary package solutions.

2.4   Misconceptions of packaging

Due to the rapid development in retail and in the production and consumption of goods, one can argue that in some parts of the world the amount of products exceeds our basic needs and in turn, the amount of packaging that naturally follows this development. But it is important to underscore that the excess is predominantly in products, since packaging seldom is present on its own, but almost always in terms of the need for products to be distributed, protected and used. Yet, this excess in products results in sub-optimized drawbacks in how consumers and authorities view packaging in their efforts to legislate its minimization. This is most likely because the package is the obvious and visible artefact in waste and landfills after product consumption.

2.4.1   Overpackaged or underpackaged?

What happens when packaging is minimized as suggested in many regulations? If the focus is set on the package alone, without considering the product inside, minimizing the packaging material can result in more product waste. Obviously when designing packaging, you need to be careful not to overuse resources by overpacking the product with an excess of packaging material or with unnecessary "double packaging". Still, it is more important to protect the contents of the package than to reduce the packaging to the point where there is insufficient protection for the product, since that will have a greater negative environmental impact in the long run, as illustrated in Figure 2.4.

The graph in Figure 2.4 shows that an increase in packaging material through overpackaging increases the negative environmental impact. But more importantly, the same graph shows that when products are underpacked, the environmental burden increases more rapidly due to the higher environmental impact caused by the combination of the wasted product and the used package. It is evident that less packaging material is used in the underpackaging compared to the overpackaging on the right side of the diagram. Yet the environmental impact is higher on the left side, when the package does not protect the product inside in a satifactory way.

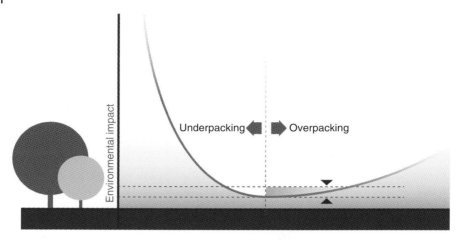

Figure 2.4 Schematic environmental impact of overpackaging and underpackaging.

By protecting and facilitating the distribution and use of products, the package contributes to sustainable development through efficient resource utilization and product protection. Sustainabiltiy thus need to be an intrinsic part of packaging design. This is why the package is instrumental in the design of products and in the assessment of sustainable solutions for sustainable supply chains. In order to make a fair assessment of the environmental impact, you need to analyse the entire system. This means that the product and its packaging system need to be assessed in the context of the entire supply chain from raw material production to recirculation or elimination. In such assessments, the environmental impact of greenhouse gases caused by different sources is often measured in terms of the carbon dioxide equivalent (CO_2e) unit. The main greenhouse gases (GHG) in the atmosphere are water vapour, carbon dioxide (CO_2), methane (CH_4), nitrous oxide (NO_X) and ozone (O_3) (Brander and Davis, 2012). For any quantity and type of greenhouse gas, CO_2e signifies the amount of CO_2 that would have the equivalent global warming impact.

Several studies have found that for food products in particular, the environmental impact of packaging is relatively small compared to that of the packed product (Nilsson et al., 2011; Wikström and Williams 2010). As already mentioned, the protective role of packaging is a key function, especially in food supply chains, since it contributes to sustainable development by less total environmental impact. As one example, a life cycle analysis (LCA) of pork showed that the production of the meat at the farm represented 89% of the total climate influence, while the package represented only 3% (Nilsson and Lindberg, 2011). In this case, an increase in packaging material that could reduce the waste of the meat content after it leaves the farm would result in a lower total climate impact from the life cycle of pork (Nilsson et al., 2011). Another example is bread, where a reduction of bread waste by 2% could motivate an increase of about 50% of bread packaging material because of the total overall reduction it would have on the environmental impact from the life cycle of bread (Williams and Wikström, 2011).

These examples show that the environmental impact of packaging is usually low in relation to the production of food (Hanssen, 1998). Normally, it is in the range of a few percent, depending on the kind of products that are produced and packed. However, examples do exist where the packaging represents up to 20% of the emissions, depending on the CO_2e relation of product and package. It also depends on the package design and the packaging material used. This means that careful considerations and assessments of the product and the package together are needed in order to develop sustainable solutions.

2.4.2 Wasteful or useful?

The role of packaging in reducing product waste that can result in a total positive environmental impact on a global scale is interesting and necessary. In Sweden alone, 100,000 tons of food are spoiled annually according to the Swedish Retail Grocers Association (Svensk Dagligvaruhandel). The average Swedish consumer wastes about 25% of the food he or she purchases. The contribution that the protective role of packaging can make to reduce food waste is thus important. Assuming that improved packaging can impact 5% of the food spoilage, it would mean that 5,000 tons of food can be consumed rather than thrown away.

Yet the common public view of packaging is the waste people see and experience in their garbage, or see as outdoor litter. This view has triggered the waste directives resulting in incentives to minimize packaging through governmental regulation. The focus on the waste aspects of the packaging supply chain has led to biased preconceptions about packaging among consumers and authorities.

When seen in waste collection, the protection role of packaging is often disregarded by neglecting to look into the entire system. This one-eyed perspective, combined with the lack of knowledge among consumers and authorities, has resulted in the focus on packaging as a great "waste" of materials. According to a study by AMR Research (2008), 76% of sustainability efforts are aimed at reducing packaging waste, where packaging is considered as a stand-alone product, and thus a stand-alone environmental burden. This view fails to differentiate between "useful" and "wasteful". Packaging that fulfils its task of ensuring safe delivery of the packaged product is useful in preventing loss of the actual goods, not wasteful. When packaging is designed to fulfil its different tasks and viewed as a system consisting of the product and the different levels of the packaging system together as a whole, it can contribute to sustainable development. From an economic point of view, balancing the cost of packaging with the cost of wasted products, as in Figure 2.5, will help to better evaluate the economic side of sustainable development.

Recently, packaging is being recognized in the literature as part of sustainable development but on a limited level due to lack of knowledge and to the traditional sub-optimized perspectives. The impact of packaging most often has focused on sub-optimized specific aspects such as returnable packaging systems or environmental-friendly packaging material, and less frequently from the holistic or systemic point of view. In the assessment of packaging, a "double" environmental impact has to be evaluated: that of resources used in the life cycle of the package itself; and that of the packages' ability to reduce the environmental

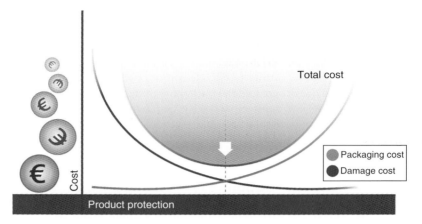

Figure 2.5 Total cost of product waste and packaging.

impact of the product it contains by protecting it from waste. The "second side of the coin" needs to be balanced with the first, which is why the packaging system and the product system need to be seen in combination.

2.4.3 How about the R's in waste hierarchies?

Previous developments for sustainable societies have led to approaches to waste thinking. These in turn have contributed to the development of concepts such as the "waste hierarchy", including the "3 R's" one should aim for when developing or using packaging:

1) Reduce
2) Reuse
3) Recycle

The first R obviously encourages us to reduce packaging in order to reduce waste. This first level of the waste hierarchy focuses only on the part of the environmental burden caused by the packaging, and not on the entire system including the product. To reduce is a good initiative for sustainable development, as long as it counts for the entire system, and if it leads to fewer products being produced. But there are initiatives under the reduce label that counteracts sustainable development. For example, consumers are encouraged to buy "big packs" in order to reduce the amount of packaging in relation to the amount of product. This counteracts sustainable development because it promotes consumers to buy more products than they actually need, and in so doing, increase the risk for more products going to waste because they are not consumed.

The 3 R's in the waste hierarchy have been expanded to the "7 R's" of sustainability by WalMar,t who uses them in a sustainability scorecard. The 7 R's include the concepts: Remove, Reduce, Reuse, Recycle(able), Renew(able), Revenue and Read (Walmart, 2014). Recently, the R's grew in number to 10, by adding and changing some of the 7 R's: Respect, Refuse, Reduce, Reuse, Renew, Recycle, Responsibility, Rethink, Replant and Restore (Earth Month Network). Once again, if these actions are sub-optimized to only one part of the system (i.e. packaging), it can counteract sustainable development. Whether 3, 7 or 10 R's, the initiative often results in promoting sustainability through the "least possible packaging" initiatives – using the smallest amount of material needed to effectively store, transport, display and sell a product. With the other perspective of the package as a contributor to waste reduction, the R's need instead to be seen from a systems perspective where the product and the entire packaging system is integrated in the term "reduce". This means that the total amount of waste from a product including its packaging needs to be reduced.

2.5 Packaging contributions to sustainable development

Any effort to achieve sustainable development or to make assessments of sustainability must apply to an entire system set by systems boundaries. In such assessments the whole packaging system has to be considered as combined, including primary packaging, secondary packaging and tertiary packaging, interconnected with the product (Azzi et al., 2012; Olsson and Larsson, 2009; Svanes et al., 2010). The packaging system further needs to be viewed throughout its entire life cycle combined with the production and distribution of goods in supply chains. The entire supply chain needs to be considered from raw material to end consumption including waste handling, rather than only assessing parts of it (Vasileiou and Morris, 2006).

With a view of sustainable development as a continuum and not a present state, we need to take a step forward in relation to the state of *sustainable packaging*. You can never lean back and be satisfied with the present state, but need to focus on *packaging design for sustainable development*, which indicates potentials and continuous future development that contribute to positive changes. To do so, new knowledge and new models for packaging design and assessments for sustainable development are needed to move new opportunities for improvements forward. However, we first need to understand and acknowledge the reasons why we have packaging and its contributions to the three pillars of sustainable development.

2.5.1 The reasons for packaging

The packaging around us is not present on its own merits and hardly ever present as a stand-alone artefact. The roles of packaging to protect products and facilitate their distribution are the main reasons for the existence of packaging. But over time the functions of packaging have diversified to also serve as a marketing and sales tool, and as a facilitator of convenience for users. This has made packaging more complex, harder to understand and to develop in sustainable ways.

In many consumer product industries, packaging has the potential to contribute to sustainable development in functions other than product protection and distribution. For example, a package has a technical role to assure quality, to safeguard hygiene and to contribute to product preservation. It has an ergonomic role to facilitate use and convenience for people, for example with easy opening and reclosing features. And it has a communicative role in being the interface for information sharing between the brand owner and the different actors that use and come in contact with it. All levels of packaging – the primary, secondary and tertiary – have information that communicates with the different users along the life cycle of the product.

2.5.2   Adding value for people, profit and planet

A product and its packaging system add value to different users, in different ways and in different situations and contexts. These values can if carefully designed contribute to all three pillars of sustainable development. One value adding aspect is the efficiency gained in handling, distribution and transport of products (Bowersox and Closs, 1996; Saghir, 2004). Package design has for example a direct impact on fill rates and thus on the transport load in supply chains (Olsson and Larsson, 2009). This indirectly affects both profit and the planet. Package design that carefully considers apportionment and unitization of the packaging system has direct impacts on consumption volumes and in that way, indirectly on product waste. Proper apportionment and unitization will also affect fill rates and thereby efficiency of transport, affecting both the planet and profit in a positive way.

The design of packaging also impacts people directly and indirectly. Easy handling does not only affect consumers but all of the employees who work with the distribution and handling of goods. Packages that are easy to carry, handle and recycle affect human sustainability because of fewer injuries, and sustainability in general because of more efficient product use. Proper apportionment contributes to people because fewer products are destroyed and wasted; it also increases product availability to more people. This indirectly affects sustainable development on the people pillar.

In the communicative role of the package in retail, the package adds value as the interface between the product and the consumer (Rundh, 2005). This is where the package provides factual information that helps consumers make conscious choices, thus affecting the sustainability of their actions directly or indirectly (Lindh et al., 2016; Olsson and Larsson, 2009).

Well thought out packaging design can contribute to economic profitability, environmental resource effectiveness and social welfare through its different functions. By evaluating packaging as a whole and not only making decisions based on its cost or price, or for that matter on only one of the three pillars of sustainability, the packaging can contribute to adding value for users, which in turn contributes to sustainable development.

All of the above-mentioned functions, and many more, need to be considered in the design of packaging systems, since they all directly or indirectly affect sustainable development. In the article *Good and Bad Packaging: Who Decides?* Robertson (1990) suggests that assessing the sustainable impact of packaging can be meaningful only if you are cognizant of the different functions of packaging and the effects they have on sustainable development.

2.6   Packaging contributions to sustainable development for supply chains

It is obvious that the majority of packages are there for the product and not as stand-alone items, since "it is almost always not products, but packed products that are handled in supply chains." The role of the packaging as the intermediate between products and different actors along the supply chains manifests its potential for contributing to all three pillars of sustainable development. In this context, sustainable development thus needs to be based not only on the flow of products, but on the flows of products *and* their packaging.

2.6.1   The research on sustainable supply chains

The research on supply chain management recognizes the need for sustainable development. Much of it concentrates on how supply chain processes need to be improved and developed to contribute to a more sustainable world. A number of concepts are used to highlight the sustainability issues, such as corporate social responsibility, green purchasing, life cycle thinking, reverse logistics, environmental logistics and circular economies. However, the research in the area usually focuses on only one at a time.

Considerable research also focuses on specific processes in two areas in particular: 1) product distribution and transport; and 2) the internal development and production processes of the products. Only a few studies integrate packaging processes, such as packaging operations or packaging development, in elaborations or assessments of the supply chain for sustainable development (Sohrabpour et al., 2016). This may be because of limited knowledge in the research community about the role of packaging or because of the misguided assumption that the package has a minimal impact on sustainable development. It is a pity that the role of packaging is so poorly understood and often neglected or disregarded in supply chain management, especially since it is such an important contributor to sustainable development.

However, some researchers argue that packaging is one of the most important areas for achieving smooth and efficient logistics operations in an international context that lead to sustainable development (Lancioni and Chandran, 1990). This acknowledgement needs to be integrated into all assessments and processes for the development of sustainable solutions. The World Economic Forum has also recognized the potential impact that packaging can have for sustainable development in supply chains. In its report entitled *Supply Chain Decarbonization* (World Economic Forum and Accenture, 2009), they assessed the possible commercial opportunities to lower the CO_2 emissions in supply chains. They found that one of the three areas with the greatest overall potential and with the highest implementation potential is new packaging initiatives. Consequently, in order to have an impact on sustainable development, supply chains would benefit from making prioritized changes in packaging design. This means that supply chain management needs to have a holistic and integrative approach that includes all the processes that make up these supply chains, including the ones for packaging.

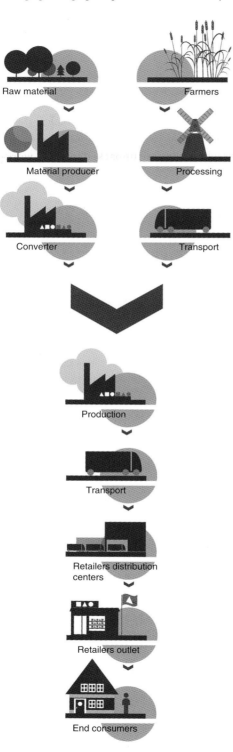

Figure 2.6 A generic food and packaging supply chain.

2.6.2 The packaging impact in retail supply chains

The retail sector is the major destination for produced products before they end up in the hands of the consumer. Developments have resulted in both producers and retailers becoming global actors, visible in several countries and on different continents. Products have to be distributed longer distances from the point of production to the point where they meet the consumer in the purchase situation (Figure 2.6). Independent of the retail set-up, almost all products sold, whether local or global, come in some kind of packaging. If not all the way to the hands of the consumer, at least to the retail stores where they are unpacked by retail workers. In both cases, the packaging is supposed to be reused, recycled and returned, or in the worst case, just wasted by the consumer (primary packaging) or the retail worker (secondary packaging).

The largest volume of packages used throughout the world is in the food retail industry, due to our everyday consumption of food (Bourlakis et al., 2011). A number of recent publications on sustainable development have specifically addressed food supply chains. But again, few have paid attention to the role of packaging. A study in the US found that the average household's climate impact related to food was about 8.1 ton CO_2e/year. Out of this, the actual delivery from the retail store to the home of the consumer (the food miles) accounted for 0.4 ton CO_2e/year. While the total freight for producing and distributing food, including the transport of raw material and other resources for production, accounted for 0.9 ton CO_2e/year. Based on these numbers, you can estimate that the transport of food production and consumption accounts for only 11% of the total emissions, wholesaling and retailing account for about 5%, and that production accounts for 83% (Weber and Mattews, 2008). The approximate 80%, accounted for in food production, supports the relevance of designing packaging systems that fulfil their role in protecting and distributing food under safe conditions to consumers. With evidence like this you need ask yourself, when will the myths about packaging turn into thoughtful packaging designs that are neither wasteful nor underpacked?

References

AMR Research (2008): http://www.amr-research.com/

Azzi A., Battini D., Persona A. and Sgarbossa F. (2012), Packaging design: General framework and research agenda. *Packaging Technology and Science*, 25, 435–456.

Beckeman M. and Olsson A. (2005), Driving forces for food packaging development in Sweden. *World Food Science*, 1–15: http://www.worldfoodscience.org/cms/

Bowersox D.J. and Closs D.J. (1996), *Logistical Management – the Integrated Supply Chain Process*, International Edition. McGraw-Hill, New York.

Bourlakis M., Vlachos I. and Zeimpekis V, (2011), *Intelligent Agrifood Chains and Networks*. John Wiley & Sons, Chichester, UK.

Brander M. and Davis G. (2012), *Greenhouse gases, CO_2, CO_2e and carbon: What do all these terms mean?* In: Ecometrica, p. 2.

Carroll A.B. and Shabana K.M. (2010), The business case for corporate social responsibility: A review of concepts, research and practice. *International Journal of Management Reviews*, 12(1), 85–105.

Earth-Month Network: http://www.earth-month.org/the-10-r-s-of-sustainability/

Elkington J. (1997), *Cannibals with Forks: The Triple Bottom Line of the 21st Century Business.* Capstone Publishing, Oxford.

Fitzpatrick L., Verghese K. and Lewis H. (2012), *Developing the Strategy.* In: *Packaging for Sustainability*, Verghese K., Lewis H. and Fitzpatrick L. (eds), Springer, London.

Hanssen O.J. (1998), Environmental impacts of product systems in a life cycle perspective: A survey of five product types based on life cycle assessments studies. *Journal of Cleaner Production*, 3–4, 299–311.

Lancioni R. and Chandran R. (1990), The role of packaging in international logistics. *International Journal of Physical Distribution and Logistics Management*, 20(8), 41–43.

Lindh H., Olsson A. and Williams H. (2016), Consumer perceptions of food packaging: contributing to or counteracting environmentally sustainable development? *Packaging Technology and Science*, 29(1), 3–23.

MBV (2007), Environmental Advisory Council, Report MVB 2007:03 (in Swedish).

McWilliams A., Siegel D.S. and Wrigth P. (2006), Corporate social responsibility: Strategic implications. *Journal of Management Studies*, 43(1), 1–18.

Nilsson K. and Lindberg U. (2011), *Klimatpåverkan i kylkedjan – från livsmedelsindustri till konsument" (Climate Effect of the Cold Food Chain – from the food industry to the consumer – the authors* translation*)*, National Food Agency, Sweden (Livsmedelsverket), Report 19-2011.

Nilsson F., Olsson A. and Wikström F. (2011), Toward sustainable goods flows – a framework from a packaging perspective. *Conference Proceedings Nofoma.*

Olsson A. and Larsson A.C. (2009), *Value Creation in PSS Design through Product and Packaging Innovation Processes*, Chapter 5. In: *Introduction to Product/Service-System Design*, Sakao and Lindahl (eds), Springer, pp. 93–108.

Robertson G.L. (1990), Good and bad packaging: Who decides? *International Journal of Physical Distribution & Logistics Management.* 20(8), 37–41.

Rundh B. (2005), The multi-faceted dimension of packaging – marketing logistic or marketing tool? *British Food Journal*, 107(9), 670–684.

Saghir M. (2004), *A platform for Packaging Logistics Development – A Systems Approach.* PhD Dissertation, Department of Design Sciences, Division of Packaging Logistics, Lund University, Lund.

Sohrabpour V., Oghazi P. and Olsson A. (2016), An improved supplier driven packaging design and development method for supply chain efficiency. *Packaging Technology and Science*, 29(3), 161–173.

Svanes E., Vold M., Moller H., Kvalvåg Pettersen M., Larsen H. and Hanssen O.J. (2010), Sustainable packaging design: A holistic methodology for packaging design. *Packaging Technology and Science*, 23(3), 161–175.

Swedish Retail Grocers Association (Svenska dagligvaruhandel) http://www.svenskdagligvaruhandel.se/

Twede D. (2012), The birth of modern packaging: Cartons, cans and bottles. *Journal of Historical Research in Marketing*, 4(2), 245–272.

UN homepage: http://www.un.org/sustainabledevelopment/sustainable-development-goals/

Vasileiou K. and Morris J. (2006), The sustainability of the supply chain for fresh potatoes in Britain. *Supply Chain Management: An International Journal*, 11(4), 317–327.

Walmart's packaging scorecard; The 7 R's of sustainable packaging: http://www.s-packaging.com/7-rs-sustainable-packaging/

Weber C.L. and Matthews H.S. (2008), Food-miles and the relative climate impacts of food choices in the United States. *Environmental Science and Technology*, 42(10), 3508–3513.

Wikström F. and Williams H. (2010), Potential environmental gains from reducing food losses through development of new packaging – a life cycle model. *Packaging Technology and Science*, 23, 403–411.

Williams H. and Wikström F. (2011), Environmental impact of packaging and food losses in a life cycle perspective: A comparative analysis of five food items. *Journal of Cleaner Production*, 19(1), 43–48.

World Economic Forum and Accenture (2009), Supply chain decarbonization – the role of logistics and transport in reducing supply chain carbon emissions: http://www3.weforum.org/docs/WEF_LT_SupplyChainDecarbonization_Report_2009.pdf

3 Designing packaging

The last decade has seen a rising interest in packaging design among scholars. As an example, Azzi et al. (2012) identified 89 original research articles that were published in this area from 1990 to 2011. These studies highlight an overwhelming range of contributions, from specific tools to management concepts. They include everything from technology and innovation design choices, to support in understanding the many facets and impacts of packaging on a wide range of topics. They cover ergonomics, logistics, sustainability and safety. This growing interest in packaging design has been followed by an increasing awareness of the benefits of well-designed packaging in the education and practice of designers. However, it is staggering that there is still such a lack of definition and cohesive description of what packaging design really is.

So what is packaging design? As we are already aware, packaging comes in an endless variety of shapes, materials, structures, colours, dimensions, configurations and imagery. These are regarded as the output of a packaging design process. This means that package design is definitely not just about the graphic design printed on the outside of packages. It is so much more than that. In short, different definitions of design emphasize different aspects of design. As Heskett (2002: 5) states, "... discussion of design is complicated by the initial problem presented by the word itself. 'Design' has so many levels of meaning that it is itself a source of confusion." According to Ravasi and Stigliani (2012), design can be perceived as an outcome, as a process, as the purpose of that process, and as the ability (or capability) to reach that purpose.

A pragmatic definition of design centres on "what it is" and "what it does". Thus, we conceive packaging design *to be a set of choices regarding the form and the function of the packaging system, as well as the activities that underpin these choices.* This means that packaging design is considered not only in terms of the final outcome, but also in terms of the process leading to that outcome.

3.1 The complexity of packaging design

That a well designed packaging system is difficult or complicated to achieve is an understatement. We argue that a more precise word describing the packaging design process would be "complex". Packaging design involves the consideration

Managing Packaging Design for Sustainable Development: A Compass for Strategic Directions, First Edition. Daniel Hellström and Annika Olsson.

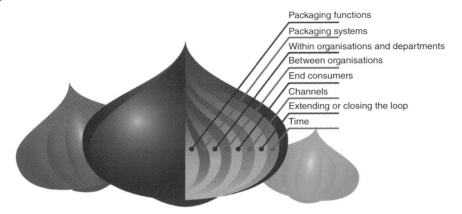

Packaging functions
Packaging systems
Within organisations and departments
Between organisations
End consumers
Channels
Extending or closing the loop
Time

Figure 3.1 Like an onion, packaging design has interrelated layers of needs and requirements.

of numerous needs, requirements and constraints; several of them often go beyond the packaging itself. The complexity of packaging design is best understood by describing the needs and requirements using the layers of an onion as a metaphor. Each layer may not be particularly complex in and of itself, but it is the combination of all the layers – the entire onion – and the requirements and interrelations of each layer that create the complexity of packaging design. Figure 3.1 illustrates the most influential layers, although the design process is not limited to these.

3.1.1   Layer one – packaging functions

The inner layer, or core, of packaging design is its functions. Understanding what packaging is intended for is essential. What is its task? What is its job? Why do we need it? The role of packaging and its functions have already been described previously. The functions tell us that the jobs and tasks that packaging has to perform are multifold. Considering all these functions is already difficult in the first layer of the design process, since the different functions are often conflicting. This requires careful and systematic trade-off decisions at this early stage. Moreover, integrating the multiple functions with the next layer of needs and requirements is usually not accomplished without "thinking twice", and that is the point where the complexity starts growing.

3.1.2   Layer two – the packaging system

Layer two involves the move from thinking in terms of the single package and its specific functions to considering its content – the product(s) – and the packaging system (primary, secondary and tertiary) as a whole. With the aim of improving the performance of the packaging system, this layer centres on the product and the packaging system as parts that together constitute the entire packaging performance. Related to layer one, which is focused on the functions of a single part, this layer considers how the packaging functions are distributed between the parts. For example, if you are able to design and stack a product in a way that assists in carrying the load-bearing strength, you will be able to use less packaging and cushioning material. Complying with all relevant legislation and regulations needs to be considered in layer two as well.

3.1.3 Layer three – functions and departments within organizations

Layer three focuses on intra-organizational aspects. By this we mean the different and sometimes conflicting needs and requirements put forth by different organizational functions and departments. Hence, various sets of competencies need to be considered: engineering, R&D, manufacturing, service, sales, marketing, finance, logistics, purchasing, legal departments, etc. They all have to be able to contribute from their packaging design perspectives. It is not always the case that departments communicate and cooperate well with each other; however, in packaging design this is needed in order to leverage from the total value of packaging design.

3.1.4 Layer four – between organizations

To make matters even more difficult, layer four takes a step further and moves from intra- to inter-organizational requirements and needs. In this step, businesses and supply chain actors often need to collaborate closely. These actors include the packaging supplier, brand owner, manufacturer, freight forwarder, wholesaler, retailer and recycler. But very often the demands from different actors are contradictory. The packaging supplier wants something that can be produced economically. The manufacturer wants something that is compatible with existing production facilities and often emphasizes that production aspects – such as line efficiency, filling speed, closing and scaling technologies – are of prime importance. The distributer wants something that enables easy handling and that is compatible with all material handling equipment. The retail store wants something easy to replenish that is also attractive to its customers. The end consumer wants something usable. Finally, the recycler wants something that is easy to reuse. The needs of all the supply chain actors differ and are often in conflict. Conflicts can also arise between the same actors: two freight forwarders, for example, may have dissimilar needs.

Outsourcing, horizontal integration and diversification are making supply chains more fragmented, adding even more complexity. Nevertheless, if the packaging is designed in a process that has representatives from all the constituencies present at the same time, it is possible to reach satisfactory solutions for all the needs, if they can manage to make decisions on how to compromise along the way. It is in the absence of cooperation, or when the disciplines operate independently, that major clashes and deficiencies occur, and poor solutions reach the market.

3.1.5 Layer five – end consumers

The dynamic diversity of requirements from the end consumers or users is a layer of its own, even though the consumer could be viewed as the final actor in a supply chain in layer four. Since there is no such thing as the average consumer, or of specific consumption situations, this poses a particular challenge for packaging designers who usually have to come up with a single design for "everyone" in "every situation". One design simply cannot work for all. Packaging often needs to be differently designed for different end consumers/users.

Non-average actors can never be neglected or overlooked, even though they are often regarded as only being the outer layer or last step in the supply chain. Even if all the previous layers have succeeded perfectly in accommodating the needs from the different actors and the requirements of the product, omitting consumers and their needs ultimately leads to unsold products and failures on the market. The consumer may come late in the supply chain, but certainly needs to come first in the design process. Having insights and understanding about consumers in terms of identifying their needs is key to starting the design process.

3.1.6   Layer six – distribution channels

"One design does not fit all" applies to more than the end consumers. Adding to the complexity, the sixth layer represents the needs and requirements of different distribution channels. A direct channel is when the brand owner or manufacturer sells directly to the consumer. An indirect channel is when the brand owner or manufacturer sells the goods to intermediaries such as retailers, agents or wholesalers. Each of these channels places certain sets of needs on packaging, depending on the abilities of the channels. Moreover, channel preferences vary between customers, and consumers are increasingly becoming multi-channel oriented, preferring different channels at different times and on different occasions. Hence, a packaging designer needs to consider a combination of channels and be multi-channelled in his or her design. The purpose is to more effectively reach different consumer segments. The main advantage is that multi-channels expand the distribution system and more customers can benefit from the products and be satisfied. Thus, packaging needs to comply with the needs from respective channels.

3.1.7   Layer seven – extending or closing the loop in circular systems

When the product has been used or consumed by the final user, or when the secondary and tertiary packaging have been used somewhere in a distribution channel, the discarded packages are about to be reused, recycled or disposed of. Layer seven emphasizes that packaging can have multiple lives where its lifecycle is extended beyond its initial use. This gives rise to additional needs in packaging design. The needs go beyond considering typical returnable packaging systems, such as the many plastic crates used in closed loop supply chains. The principle of multiple life packaging is to reuse packaging in innovative and creative ways. There are a great number of examples where packaging remains in people's homes or workplaces and serves new purposes. In many of these situations the reuse is closely related to brand perception, and the more useful and unique a design, the greater the chance of the package being reused. Hence, designers need to consider giving packaging additional life before it is recycled.

Layer seven also involves closing the loop so that it forms circular systems. This can be achieved by transforming the packaging into a completely different object. Obviously, this also includes viewing the recycled packaging material as a "secondary raw material", which is an increasingly common initiative found in the consumer packaging considerations of the brand owner. Turning packaging into energy is another way of closing the loop.

One could argue that this layer – extending or closing the loop in circular systems – is part of previous layers. Layer four also includes the recyclers' needs, which are seldom involved or considered in the design process. Likewise, layer five considers the end consumers' reuse, recycling and disposal behaviours of packages. But since consumers behave irrationally, the recycling layer needs special attention. Bear in mind that all layers are interrelated: layer seven adds to the previous layers a sort of iteration, or more importantly, extends additional loops in time. That brings us to the next layer.

3.1.8   Layer eight – time

And finally – all the different layers are dynamic! New materials, methods, technologies and market opportunities are continuously changing and evolving. Likewise, global issues, such as the changing social, cultural, political and demographic environments that shape our complex world, all affect packaging design. In the onion-layer metaphor of the needs and requirements in the packaging design process, you could say that the onion is growing or decaying, and that means that its needs and requirements are constantly in transformation. It is important to be aware of trends and developments on the global, regional and local scale.

3.2   Challenges of dealing with the complexity

Packaging design is certainly complex and is constantly changing. Accordingly, we need to pay attention to a wide range of needs, requirements and constraints. Unfortunately, this complexity makes it practically impossible to design the optimal packaging system because one design does not fit all. At the same time, it presents an excellent opportunity for those who are looking to make creative and innovative solutions, since the complexity is a major reason why there are always opportunities for improvements in packaging design. That is also why it is such a rich, engaging profession with results that can be very impressive, especially in the way they can contribute to sustainable development. We find packaging design to be rewarding. On the one hand, it offers opportunities to develop packaging that assists and enriches the lives of people, benefits businesses, and makes the planet a better place; but with distressingly complex constraints to overcome on the other.

So, complexity is not bad if you learn how to embrace and overcome it. The most important principal for taming complexity is to give structure and understanding to the design process itself. Complex things are no longer complicated once they are understood. To satisfy the myriad of needs, requirements and constraints of packaging necessitates multifaceted skills and a large portion of patience. What is required is a combination of high technological skills, great business skills, and well-developed personal and social skills to interact with many different groups of professions, all of whom have their own agenda and believe their requirements are the most critical. But that is not enough. Packaging design also requires great management, because perhaps the hardest part is coordinating all the many, separate disciplines and organizations, each with different goals and priorities.

Each profession, department and organization has a different perspective of the relative importance of the many factors that make up the packaging system. One department argues that it must be logistically and environmentally efficient, another that it must be consumer convenient, and yet another that it has to be low cost. Managers are themselves key stakeholders in dealing with the complexity of packaging design and in setting the overall strategy for the outcome.

To deal with the complexity of packaging design, managers and other decision-makers have four major, interconnected challenges to address. Depending on the complexity, they vary in magnitude. The packaging design challenges are:

1) taking a holistic approach;
2) integrating form and function;
3) making trade-off decisions; and
4) sharing the risks and gains.

We view these as a series of wonderful and exciting challenges, each of which can be seen as an opportunity.

3.2.1   Taking a holistic approach to packaging

The first and foremost challenge managers need to address is that of taking a holistic approach. Even though this has been called for by a long list of researchers (for example, Colwill et al., 2012; Hellström and Saghir, 2007; Jahre and Hatteland, 2004; Johnsson, 1998; Molina-Besch and Pålsson, 2016), you may wonder what it really means. By holistic, we mean placing the emphasis on the whole rather than its separate parts. When we apply this concept to packaging design, it makes us aware that we cannot really carry it out without knowing and learning about all the needs (the tasks, the functions, the stakeholders involved, the conditions to be aware of, etc.) and understanding the impact of the interrelations between these needs. However, "holistic" is in a sense a relative concept. There are those who have a "limited" holistic approach based on a certain point of view. Examples would be designing packaging systems strictly based on the needs of the end consumer in order to maximize brand owner profit, or that are strictly based on distribution needs in order to maximize handling efficiency. On the contrary, these perspectives are only a fraction of a holistic approach, since the entirety is not emphasized in either.

The aim of attaining a holistic approach is to enable rigorous packaging design decisions that consciously consider the complexity involved. These decisions may be directed towards improving overall supply chain efficiency, limiting the overall cost, launching a new product in an emerging category and/or market, or reducing the social and environmental impact. Thus, a clear overall strategy is needed for a holistic approach. And what better strategy is there than that of sustainable development? As sustainable development is centred on the planet Earth, it is a good strategy for a holistic approach because it captures all the relevant perspectives on packaging both in space and time. The holistic approach to packaging design brings together the disparate aspects of packaging complexity so that they support each other and serve as a backbone for integrating social, economic and environmental development. One can view this aim of attaining a holistic approach as utopian. Nevertheless, it is the foremost challenge in managing packaging design.

Embracing a holistic approach to packaging design is not straightforward. Moreover, packaging designers and managers think and reason differently. One way would be to peel the onion layer by layer, understanding and considering one layer at a time, but with the overall strategy in mind. Every contrasting layer and its interconnections can and must be clearly understood before you can truly design packaging systems from a holistic approach. Often packaging design processes begin at the inner layers, and then at best move outwards. However, beginning in the opposite direction, at the outer layer and moving inwards, would also work. You could also consider an "onion soup" way of thinking, where the performance of a packaging system is the result of an interwoven network of aspects, often emerging and unknown in the system, but potentially valuable for the stakeholders and organizations affected. No matter which way of thinking is used, the next challenge needs to be considered, that of combining form and function in the design process of the product and packaging systems.

3.2.2 Integrating form and function

The second challenge is heavily related to the previous one and concerns the combining of form and function in the design process of the product and packaging system. As you already know, packaging is a necessity for almost every type of produced artefact that is to be transported from the place of production to the place of consumption. It does not matter if the artefacts are mined, grown, extracted or manufactured, packages are needed for them all. For consumables, the packaging is most often regarded as an integrated part of the product and thus seldom noticed by users as an artefact on its own. As a result, it is also often forgotten. This is the case not only for the end users who naturally regard the package and product as integrated, but also for the product developers and producers. They have traditionally viewed packaging as something of less importance, which is why packaging design often starts only when the core product is ready to be launched (Jönson, 1993). All too often when product designers concentrate with tunnel vision on their product features, production processes and cost calculations, packaging becomes an afterthought, if even that. When it is thought of, it is in most cases way too late. And when it is too late, you either take an existing package in your portfolio, or you ask your supplier for something that already exists, since the shortage of time does not allow for new packaging development. That means that the package's form and function are designed for previous products or already existing customers or markets.

In order not to forget packaging, we have to realize that developing a product and designing its packaging are really two aspects of one single process

(Harckham, 1989). The extensive effort put into developing a product will be wasted if we neglect to consider that it needs to reach the end user in a sound and safe condition. We will have a better chance to succeed in doing so if we design packaging forms and functions that fit the requirements of the new product. The more the package and the product can be combined in form and function in the design process, the greater the probability of them being effective and efficient (Harckham, 1989). This is the foremost way to ensure that the final user receives the product in a good, satisfactory condition.

Depending on which design project, organisation or people are involved, the challenge of integrating product and packaging form and function poses varying levels of difficulty. The challenge does not imply a 100% integration, nor is this always possible. In a design there is often a need to consider multiple forms and functions and very frequently, conflicting ones. This leads to the next challenge: understanding and making trade-off decisions.

3.2.3   Making trade-off decisions

Packaging decisions are easy for some people. If you want a package to be as cheap as possible, you just buy it from the supplier offering the lowest price. You have only one, single and straightforward objective, so you only need to make a single set of comparisons. But most of us want more, such as value-based branding, consumer convenience, ergonomic and environmental friendly solutions. We also want the product to be protected and distributed efficiently. This means that the packaging decision is considerably more difficult than just comparing prices. The different, multiple and conflicting needs and requirements from a variety of stakeholders and organizations along the distribution channels that impose constraints on the packaging system make perfect decisions that satisfy all to a "mission impossible". We have to make trade-offs.

Making wise trade-off decisions is one of the most important and difficult challenges in packaging design. The sheer volume of different trade-offs makes it hard. In addition, each need or requirement has its own basis of comparison, such as quality, cost and time. And as if this was not enough, they also have a plethora of measurement scales, such as numbers, relational judgments, descriptive words and colours. This means that you are not just trading off one kind of apple for another; you are trading apples for pears or even for coconuts. Adding the fact that the different objectives from a variety of stakeholders are often conflicting makes it even more problematic. For example, a logistics manager would perceive packaging design as a pure cost-value trade-off (Lambert et al., 1998), while there are other trade-offs for production, sales, marketing and CSR managers. There is seldom a clear picture of all the alternatives and their consequences for each objective when you start a design process. To top it off, all the above trade-offs can be considered and assessed differently by the stakeholders and organizations involved.

It is important to recognize that every time a packaging design decision is made, you also have to make trade-offs. It is naive to think that successful packaging solutions have to satisfy all requirements 100% from the viewpoints of the users, the environment and the businesses. Packaging design decisions always involve compromises. If you think you have found the optimal solution, you are sadly mistaken. In the best of cases, you may only have the best alternative solution so far. Making trade-offs is and will always be an essential part of packaging design, and thus a challenge for all who are involved in the design process. Finding synergies, in terms of "win-wins" among the trade-offs is important for everyone. There are many packaging designs that capture positive interactions and synergies that drives diverse benefits. A major task for packaging designers is to determine which of the trade-offs encourage a best-fit solution with sufficient benefits that can drive sustainable development.

3.2.4 Sharing the risks and gains

Based on the explanation of the trade-off challenge, decision-makers are faced with a fourth and interrelated challenge: sharing the risks and gains. In today's businesses this is neither easy nor common. Take cost for example. The argument that the packaging design must satisfy the actor who pays for it – who is not always the primary or end user – seems logical at first glance. Most types of primary packages, such as those for food, are often purchased by the manufacturing or purchasing departments. Without a holistic approach, the purchasers in these cases will probably be most interested in the price or the process compatibility, but almost certainly not in ergonomics or the best product protection. The argument that the packaging design must satisfy the one who pays for it, is not wrong *per se*, but needs to be amended and expanded to encompass a holistic approach.

The reasoning behind applying a holistic approach to packaging design is that it will increase the total gain. But it will also increase the risk for failure because the approach crosses functions and organizational borders. Collaboration efforts represent a key factor for adopting a holistic approach where trade-offs are consciously made. According to Lambert and Cooper (2000), a critical aspect in supply chain collaboration is that of risk and gain sharing among organizations. Consequently, it is a challenge in packaging design.

"Risk" and "gain" are vague terms with different definitions. What they mean in the context of packaging design needs to be further explained. Jüttner et al. (2003) divides "risk" into "risk source" and "risk consequence/impact". Risk source relates to variables that cannot be predicted with certainty and that have an impact on outcome variables. Risk consequences/impacts are outcome variables like cost, quality, safety and health (i.e. the different forms in which the variance becomes manifest). The latter – risk consequences/impacts – is what we refer to as risk in this book, because it emphasizes that a risk is linked to a loss, a negative outcome of some kind. The term "gain" is easier to explain because it is the opposite of risk, and thus linked to a positive outcome of some kind.

In supply chain management, risk and gain sharing aspects are critical success factors (Ballou et al., 2000; Cachon, 2003; Lambert and Cooper, 2000; Mentzer et al., 2001). Narayanan and Raman (2004: 96) state that a supply chain works well if "...its companies' incentives are aligned – that is, if the risks, costs and rewards of doing business are distributed fairly across the network."

This is supported by many others, such as Lee (2004) who argues that the lack of alignment between incentives has caused several supply chain practices to fail. One way of aligning the supply partners' interests is to define the terms of their relationships so that the firms share risks and gains equitably. From a packaging viewpoint, this means that if incentives are not aligned between stakeholders, organizations and other partners involved, they will adopt unbeneficial behaviours and risk attitudes resulting in poor, sub-optimized solutions far from the preferred sustainable solutions.

Professionals involved in packaging design should expect that incentive problems and opportunities will arise when any change initiatives are implemented in the packaging system. These changes can be incremental and small in nature or sudden and major. Examples of major change initiatives would be adopting collaborative packaging guidelines, implementing returnable packaging systems, developing new standards, or adopting new, disruptive technologies such as RFID (Hellström et al., 2011). In order to make these initiatives work properly, decision-makers have to identify the incentive problems that the initiative gives rise to and take collaborative measures towards aligning the incentives. The fundamental step towards aligning incentives is to identify the risks and gains for each involved stakeholder and organization at the start of the design process.

In major packaging design initiatives, aligning the risks and gains between stakeholders may result in new business strategies or new business models. Osterwalder (2004) reports that the business model can be regarded as a translation of the strategy of the firm into a tangible model of how the firm plans to create value. The business model can, in turn, be seen as an abstraction of the business logic of the firm and be translated into certain business processes and infrastructures to fulfil the purpose of the firm. In the context of supply chains, the different business models (of the involved actors) need to be aligned in order to create value in the entire system.

Postponement is an example of a business strategy that aims to maximize possible gains and minimize risks by delaying investment into a product or service until the last possible moment. Packaging postponement is a typical type of postponement, and Twede et al. (2000) describe five factors that favour this strategy and clearly go beyond packaging itself. These factors are:

1) modular products that can be customized for local markets;
2) products that gain volume, weight or value from packing;
3) unpredictable demand;
4) a large number of market-based variations for a single formulation; and
5) and situations where economies of scale in packaging and logistics can be found.

New business models need to be implemented to lock into the agenda of sustainable development and the challenges to end poverty, protect the planet and ensure prosperity. In packaging design, the challenge to decrease product waste, for example, would mean that the organizations involved need to share the risks and gains from less production, less waste, fewer transports and fewer claims.

3.3 Organizing and managing packaging design

Good design has its origin in carefully managed processes (Bruce and Bessant, 2002; Chiva and Alegre, 2009; García-Arca and Prado, 2008). Package design can take place within a company or by utilizing various degrees of external packaging resources, such as independent contractors, consultants, vendor evaluations, independent laboratories, contract packagers and outsourcing. All of this requires some sort of project planning and project management methodology. Either way, we will discuss two basic aspects of putting the packaging design process into practice: first, the process itself; and second, managing the team that will carry out the cooperative efforts of the multiple disciplines. These two aspects are cornerstones in making packaging design a harmonious, smoothly functioning, cooperative and respectful project.

3.3.1 Design thinking processes

Based on the insights derived from the previous chapters, you can now deduce that modern packaging design processes all have the following five common, fundamental and interwoven characteristics:

1) They attempt to capture and consider the complexity of packaging design.
2) They regard sustainability as an intrinsic strategy of packaging design.
3) They integrate the form and function of products and packaging systems.
4) They seek synergies when making trade-off decisions.
5) They align stakeholder incentives with risk and gain sharing.

In order to structure the challenging characteristics mentioned above, many researchers and practitioners have proposed a variety of development processes set up as sequential activities (see for example, Bramklev, 2009; Ulrich and Eppinger, 2007; Vazquez et al., 2003). These activities need to be carried out in the given order to launch new products on the market. Most organizations have adopted sequential thinking and organized their product development into stage-gate models where a decision has to be made at each stage before the process is released into the next stage. All kinds of processes that are set up in this way help structure the design process and make it understandable to employees, but often this also locks the process and creates obstacles that can result in non-holistic approaches and a lack of well-balanced, trade-off decisions.

However, experience tells us that the initial design process is at its best when it is allowed to work iteratively and to embrace the complexity of packaging, rather than being hindered or locked into sequential stages. "Design Thinking" is one such iterative process. It was introduced in 2005 at the Hasso Plattner Institute of Design at Stanford University as a way of working with design processes for all kinds of products and services. Design thinking arises at the intersection of three major concepts that are part of the input in the design process (Plattner et al., 2011):

1) the desirability of the user;
2) the products' technical feasibility; and
3) the economic, environmental and social viability.

Transferring design thinking to packaging design for sustainable development calls for a broadening of the last concept, economic viability, to include also environmental and social viability. Consequently, applying the concepts of design thinking to the packaging design process means that you start by understanding the needs and requirements in order to find solutions that are desired by users. At the same time the designers need to consider constraints in order to evaluate whether the solutions are technically feasible to develop and produce. Finally, the designers need to consider whether the product is viable from a sustainable development viewpoint.

3.3.1.1   Desirability by the user

Users can be both end users and direct customers along the supply chain. This means that several different users are in contact with packaging. They often have conflicting demands on what a package has to perform. Whatever the needs, they are based on different user preferences that are personal, situational or even socio-economic (Grunert et al., 2010). For a packaging designer to identify what users really want or need is certainly not easy. The most intriguing part is the fact that the majority of consumers do not really know what they want from new innovative products, since they are usually constrained by their own perceptions of existing products on the market. These constraints usually result in them only being able to suggest incremental changes to existing products when asked to express their needs. For more radical or innovative products, users normally have latent needs that come into their minds once the products reach the market. Similar to user-oriented innovation or user-oriented product development processes, a good starting point in packaging design thinking is to identify the needs of different users around the package and along the supply chain through which the packaged product is to travel. Methods and tools are available for packaging designers to access user insights and needs. These are briefly explained in section 3.4.

3.3.1.2   Technical feasibility

Design thinking can be used in packaging design to consider whether the solution will be technically feasible for the product it is supposed to protect. The technical feasibility of the package has two major thoughts behind it. First, whether the requirements that the actual product places on the package are fulfilled by the proposed design, for example, if it should be shock resistant or if it should protect the product from the surrounding environment. Second, if it is technically possible to produce the package or distribute it with the required protection. This means that both the surrounding environment and the product inside place constraints on the package from a technical viewpoint. Defining the specifications of the packaging system based on technical requirements is a difficult part of the design process; so much so that some design principles avoid specifying the problem as long as possible and instead repeatedly redo rough estimates and rapid tests of ideas to improve existing solutions. Even though this task is not easy and often very time-consuming, it is worthwhile to carefully consider the trade-offs between technical constraints in order to come up with suitable packaging designs. For such exercises there are assessment methods and tools at hand, some of which are briefly explained in section 3.4.

3.3.1.3 Economic, environmental and social viability

The third aspect of design thinking is to carry out an evaluation of the economic, environmental and social viability, or taken together, the sustainable viability. The assessment evaluates the entire packaging system contribution to meeting the challenges of all three pillars of sustainable development. Sustainable viability thus needs to assess whether the entire packaging system is affordable for the people who are going to pay for it and economically sustainable for the business that produces it. The entire packaging system also needs to be assessed to determine the benefits it generates for the environmental impact throughout its entire life cycle. Finally, the social impact the packaging system has on users in the entire network has to be assessed. Some researchers have presented different frameworks for sustainable packaging development, but they are usually very complex and seldom take a holistic approach to assessing the packaging design from a sustainable viable systems perspective.

3.3.2 Managing the team

Packaging design is a creative and knowledge-intensive practice. Packaging designers are sometimes asked to figure out how to manage the complex system of packaging design with all the interactions of technology, processes, policies, people and organizations. How can package designers or anyone else for that matter, work across so many different domains? But the designers are only one of the professions involved. Packaging design is a process that requires the cooperative efforts of multiple disciplines and of different organizational functions. The number of disciplines and stakeholders that are required to design and produce a successful packaging system is staggering. A successful packaging solution requires great engineers, R&D professionals, marketing people and good managers.

In properly run organizations, team members from all the stakeholders with their different perspectives on packaging come together to share their requirements and work harmoniously to design a packaging system that satisfies the "whole" system, not just themselves, or at least that does so with acceptable compromises and trade-offs in mind for a better whole. In dysfunctional organizations, teams work in isolation, often doubting and arguing with other teams about specifications and requirements. This results in narrow-minded packaging solutions that offer limited value and in the worst case, sub-optimizations that lead to fewer benefits for all the actors and eventually, to dissatisfied customers.

A good packaging design is the result of teamwork. The design of a packaging system set-up is, of course, based on input from policy guidelines, legislation, product characteristics, the distribution set-up and marketing criteria, but the process itself is a team effort. The team members need to connect and jointly share the holistic packaging view and transform it into functions and features that fulfil the needs placed on the packaging system. Both researchers and practitioners acknowledge this. Twede (1992), for example, clearly stated that packaging design and innovation is a team effort that requires input from several company functions. From the practitioner viewpoint the statement by a global

R&D director of a food company put it this way: "Packaging is in the middle of things; all departments relate to it. This makes packaging projects attractive since you need to work in cross-functional teams."

A project can be organized and structured in many ways depending on the organizational structure (such as functional or matrix). All organizational structures have strengths and weaknesses. A functional structure has the advantage of clear lines of authority, and allows its team members to concentrate on their particular mission. The drawback is that projects can end up with members who are working towards their own functional goals and do not communicate or cooperate with each other very well. A matrix structure is more complicated and requires more administration, communication and coordination. Its strengths are in bringing together employees and managers across functional departments, which generates resource and information efficiency because experts and equipment can be shared across projects. Consequently, the project structure that you adopt depends on the organization and its context. The core is to build multidisciplinary and multifunctional integrated teams. To do this, we argue that no "radical surgery" is needed in organizational processes or re-engineering. However, getting these teams to function efficiently in practice and not end up dysfunctional is another matter. Here are a set of skills for managing and participating in packaging design:

- Appreciate the complexity of packaging design and the context in which it occurs, so that you can truly grasp a holistic approach.
- Understand how packaging designs affects the stakeholders and organizations in all aspects – social, economic and environmental.
- Establish an effective and trusting relationship with the stakeholders and organizations in order to explore their needs and options with them so that they can participate, resolve and propose packaging design ideas and solutions.
- Liaise with other significant stakeholders and organizations to facilitate and encourage them to participate in packaging design.

Determine the knowledge and insights that organizations have and convert this into useful information for the development of packaging:

- Be aware of the potential trade-offs and make decisions with acceptable compromises. Enable risk and gain sharing among those affected.
- Reach consensus among all involved parties when making decisions on sharing risks and gains.

Putting a holistic approach into practice in packaging design is far from simple and this list of skills is incomplete. However, we argue that the work of managers in packaging design is not unique and can be described as everyday managerial work: complex, dynamic and situation-dependent. It can be compared to conducting an orchestra where no rehearsals are allowed (Mintzberg, 1973), but also as being a puppet in a marionette theatre (Carlson, 1991). One thing that is certain is that team members need different kinds of competencies, methods and tools in the design process.

3.4   Tools for packaging design

The design of packaging is an act of craftsmanship and teamwork. It is characterized by the ability to employ problem-solving, creativity and decision-making in order to reach an adequate solution. Creativity is the "fuel" in packaging design projects. Irrespective of the team involved, and the tools and methods that support the design team, it will always be the people who provide and ignite the creative sparks that drive progress in packaging design projects. This is a fact. And it is supported by the packaging design process itself. This process consists of a set of activities, the results of which are unknown in advance and are achieved in unpredictable ways. Decision-making arises from the need to select the best possible course of action where the exploration, generation and comparisons of alternatives are an integral and iterative part.

Creativity and decision-making places packaging design in a bigger picture, one of strategy and management. The link between packaging design and management is represented in practice by different design methodologies and systematic approaches such as: different stage gate product development processes (PDPs) (Ulrich and Eppinger, 2007); Quality Function Deployment (QFD) (Mizuno, 1994); Six Sigma (Harry and Schroeder, 2005); and Total Quality Management, (TQM) (Martínez-Lorente et al., 1998). It is of great importance to acknowledge that the balance of creativity and systematic approaches in packaging design strongly depends on the type of project. The difference between radical and incremental innovation as well as between routine or non-routine design plays a crucial role.

The use of tools in packaging design projects is coupled with the balance between open-minded creativity and iterative and structured decision-making. Tools can stimulate creativity and also significantly improve design efficiency and the way in which the project is executed. However, the use of tools is also directly coupled to the design team activities in the design process. A salient feature of designers' expertise is to select which activity to address and the tool with which to do so. It is only in the hands of these professionals that the right tool becomes powerful.

The list of tools that can be used in packaging design projects is extensive and many organizations create and develop their own. As the use of tools is clearly connected with the activities in packaging design projects, the tools can be separated according to the "diverging" and "converging" efforts of designers. In simple terms, divergence is the first phase of a design process where the ideas or topics are expanded and where tools for capturing user insights and generating ideas are used. Convergence is the second phase, where the effort is to narrow down the considerations and solutions using various evaluation and assessment tools.

What is presented here is just a small selection of commonly used tools in packaging design projects. The descriptions are not meant to be comprehensive, but are rather short to provide an overview. One needs to keep in mind that tool selection in the design process does not depend on the tool alone (such as ones that determine functionality, cost, time to execute, implementation time, training, quality). It also depends on the characteristics of the project.

Aspects that need to be considered before choosing suitable tools are the project goal, the design phase, the team composition, different constraints, and strategic directions (Lutters et al., 2014).

3.4.1 Divergent phase tools – identifying needs and generating ideas

Creativity and idea generation are at the core of the divergent phase. This is where you can go "out-of-the-box" and beyond the boundaries that are already set. This is where you can encourage freedom of ideas and acknowledge diversity in the team (Reid et al., 2014). One challenge for design teams is to remain in the open-minded, divergent phase and trust the process of staying there long enough. It is key to spend time and resources on exploring the potential users and gaining insights about them and their situations, rather than striving to rapidly come up with a solution. For the team to rely on insight mining, idea generation and prototype iterations and the resulting creativity is time well invested. Even though the divergent phase is supposed to be open and free, there are several tools to use in the insight mining phase and in the creative phase of analysis, idea generation and prototype creation. Some of these tools are briefly explained to serve as an inspiration in packaging design projects.

3.4.1.1 User insight mining

Insight mining is a first step in a packaging design process. This is because knowing the users is crucial to designing products that fulfil their needs. Insight mining can either be done thoughtfully and thoroughly by spending time on gaining deep and rich insights about users in close proximity to the potential product use. Or it can be done by collecting considerable amounts of data in a less time-consuming manner, where the data from many distant users can be gathered and included. It is up to the design team of clarify and decide the purpose of the design project. Is it about a new radical design? Or is it about incremental changes of an existing design? This needs to be sorted out before you can determine the right insight mining tools to use. The distance of the user from the potential object will have an impact on the reliability of the insights. As seen in Figure 3.2, there are quite a few ways to acquire user insights and it is worthwhile to spend time and energy on doing so in the early phase of the design process.

3.4.1.2 Ethnographic studies – observations

What people say often differs from what they actually do. Ethnographic studies or observations can thus provide designers with insights into the potential use and obstacles that users can encounter. In-context observations are when you observe people in their real-life settings, and in their daily, normal use of products or services. Observations of this kind put you as a designer in the user's situation, and provide you with rich insights and valuable input for your design process (IDEO, 2014; Osterwalder et al., 2014). When making observations, it is important to free yourself from your own pre-set assumptions and let yourself just observe what you see and experience. Look for things that shift people's behaviour, things that people care about, things that surprise you or question your assumptions and appear irrational. Look at peoples' body language and the way they use products (IDEO, 2014).

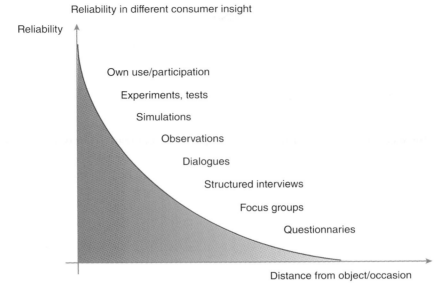

Reliability in different consumer insight

Reliability

Own use/participation

Experiments, tests

Simulations

Observations

Dialogues

Structured interviews

Focus groups

Questionnaries

Distance from object/occasion

Figure 3.2 Reliability of different consumer insight methods.

3.4.1.3 User interviews or conversations

User interviews or conversations should usually last about an hour to gain deep insights about the subject of interest and to allow the interviewee the time needed to express what he or she has to say. It is important to use open-ended questions to get input that is not biased or limited by the questions you ask, and that opens up for elaborative answers. The aim is to explore the values, frustrations, desires and aspirations of a potential user. It is important to focus on the interviewees and let them speak and feel that the conversation is about them. Listen and be attentive, and help the conversation along by asking the same questions back and forth in order to get elaborations from different perspectives.

3.4.1.4 Focus group interview

A focus group study is a qualitative method for gaining insights about a product or service from a group of users. The focus group is led by a person from the design team who guides the discussion based on a predefined guide with questions and issues. The group is asked for their opinions on a given issue or product, and are allowed to move away from the pre-set guide. A focus group will typically consist of 6 to 12 people (Cross, 2008).

3.4.1.5 Survey questionnaires

Questionnaires are most often sent out as a survey to a large group of consumers or potential users of a product or service. The advantage of questionnaires is the low cost and wide spread of consumers that can be reached in fairly easy ways. Questionnaires are most often set up as quantitative, multiple choice questions with fixed response alternatives (closed-ended question), usually based on scales with different units. For example, a scale from "really dislike" to "like very much". Questionnaires can also be constructed to have open-ended questions generating free text answers to gain qualitative input, but they require more analysis afterwards (Cross, 2008).

3.4.1.6 User participation

Instead of observing, interviewing or using other methods for gathering user insights, one can utilize the user in the design process as the designer of products. By providing easy-to-use design tools, one can enable users to develop new product innovations for themselves (von Hippel and Katz, 2002). Finding users and allowing them to develop products that can be producible custom products means moving parts of the design process from the supplier to the user themselves (Thomke and von Hippel, 2002). In that way users can create a preliminary design, simulate or prototype it, evaluate its functionality in their own use environment, and then iteratively improve it until they are satisfied. Many companies provide such easy-to-design tools in open innovation platforms on their websites.

3.4.1.7 Benchmarking

Benchmarking is a tool for gathering insights into what is already present on the market. That means checking what competitors or other industries are doing at the moment within the given product segment. Benchmarking can be carried out for different purposes, such as to gain inspiration for new ideas, but also to ensure that you are not developing something that is already out there.

3.4.1.8 Ideation

After having collected input about the potential users, it is worthwhile to spend time on generating ideas. There are several tools used by designers for this purpose. The following examples of tools are used to acknowledge several optional ideas before you focus in on one or a few potential final solutions to be taken further in the design process where they will be evaluated and assessed.

3.4.1.9 Brainstorming

Brainstorming is the most well-known and widely used method for generating a large number of ideas. Brainstorming is normally carried out in a small group such as the design team. The method benefits from having multidisciplinary competencies and diverse perspectives represented in the group and it should work in a non-hierarchical manner. One person needs to lead the session, to ensure that the format and guidelines are followed. The leader starts the session off by describing the problem. After a brainstorming session the ideas are grouped into themes established by the group based on the ideas that have come up. Here are some guidelines to follow when carrying out a brainstorming session:

- All ideas are welcome, the more the better.
- Ideas should be kept short and catchy.
- No idea is stupid.
- Crazy and out-of-the-box ideas are sought after.
- Try to combine and improve the ideas of others.

3.4.1.10 Personas

Based on the insights gained in the insight mining, a fictional character is built up that represents a typical user (Miaskiewicz and Kozar, 2011). This fictional character is called a persona and is built on a reliable and realistic representation of the target user (usability.gov, 2016). A persona can be built up as a collage of pictures or as a story that tells about the persona. It can also be presented as a combination of a story and a set of pictures.

3.4.1.11 Simulate the voice of the consumer

Use people to role play the consumer in order to bring the voice of the customer into the room of the design team and the other stakeholders (Osterwalder et al., 2014). The voice of the customer can be based on the insights from the insight mining methods used in the early phases.

3.4.1.12 Empathy maps

An empathy map is a tool for visualizing and creating a better understanding of the environments, behaviours, concerns and aspirations of the customer. When making an empathy map, you bring in the different senses of the potential user and map what that potential consumer sees, feels, hears or says and does. The empathy map is then completed by identifying potential pain points (i.e. potential fears, frustrations, obstacles) and potential gain points (i.e. wants and needs, measures of success) (Osterwalder and Pigneur, 2010).

3.4.1.13 Prototyping

Prototypes are rough conceptual models of reality that could be a potential product or service. A prototype is built in order to explain design ideas in a visual and three-dimensional way. In design thinking, the making of rapid (meaning very early in the design process) prototypes is a tool that allows people to gain an early understanding of the ideas behind your design project. Rapid prototyping allows for several evaluations and iterations of the design solution. That is why prototypes should neither be ready models nor perfect (Plattner et al., 2015). Experiencing your own prototype will give you immediate feedback on what may work and what will not. To show prototypes to potential users can help to gain additional insights into your creative and iterative design process. Showing prototypes to managers communicates the potential solution and facilitates decision-making in the design process.

3.4.2 Convergent phase tools – decision-making support

With an inventory of user insights and creative ideas, the convergent phase is where you start narrowing down your options to a few preferred choices. To eliminate less attractive possibilities and to choose a way forward in a project you can use basic logic, argumentation or reasoning to arrive at a selection of possible solutions. However, most often some kind of analysis is needed for these decisions. Here we present examples of evaluation and assessment tools. Some are general design tools or assessment methods, while others are specifically tailored to packaging design projects.

The benefit of many of these tools is often to establish a baseline or benchmark. For example, before embarking on a redesign, many designers first assess the performance of their current packaging solution to use as a baseline. These tools are used by manufacturers, brand owners, distributors and retailers, since they are responsible for regularly reviewing their packaging designs. Measuring and analysing the level of impact packaging has on the environment or on costs is fundamental, in order to consider the latest developments in both technology and society.

3.4.3 Packaging evaluation and assessment

3.4.3.1 KANO attractive quality model

KANO is a model to analyse and classify user requirements or features into five categories, in order to prioritise development efforts. It is based on theories of attractive quality (Kano et al., 1984). Applied to packaging design, KANO puts customer satisfaction in relation to the degree of achievement fulfilled by different packaging functions and features (Witell and Löfgren, 2005). Its aim is to understand how users perceive quality attributes. They are divided into five categories: 1) attractive quality; 2) must-be quality; 3) reverse quality; 4) one-dimensional quality; and 5) indifferent quality, as seen in Figure 3.3. The different qualities are evaluated against the users' level of satisfaction.

3.4.3.2 Packaging scorecard

The packaging scorecard was developed by Olsmats and Dominic (2003) and allows designers to measure the performance of a packaging system using approximately 20 functional packaging criteria. The method enables the identification of criteria that are important for respondents, any actor or users and the satisfaction of the respondents with a particular packaging system. The importance of each criterion is rated by respondents on a scale of 0–100%. This weight is then normalized, indicating the relative significance of each criterion. The respondents also evaluate the criterion fulfilment by assigning a score between 0 (not applicable) and 4 (excellent performance). Multiplying the score by the normalized weight gives an indication of how the package is performing. Because the data is collected from either structured interviews or questionnaires, it is for obvious reasons of great importance that the respondents have knowledge and understanding of the packaging system. Consequently, the packaging scorecard can identify gaps between the perceived importance and satisfaction of different packaging criteria. Pålsson and Hellström (2016), for example, used the packaging scorecard to assess a packaging system's performance in order to explain the current state, trade-offs and the packaging related improvement potential in supply chains.

3.4.4 Environmental evaluation and assessment

3.4.4.1 Life Cycle Assessment (LCA)

"Life cycle" is defined in the ISO standards (ISO, 14040, 2006) as: "consecutive and interlinked stages of a product system, from raw material acquisition or generation from natural resources to final disposal." This indicates a "cradle-to-grave"

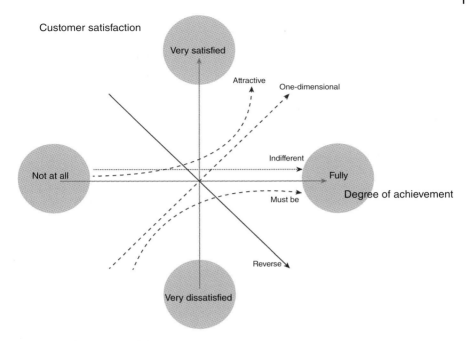

Figure 3.3 The KANO model.

view of product life cycles. An LCA is used to make an environmental assessment of a given system within the system's cradle-to-grave boundaries. An LCA involves: "…careful compilation and evaluation of the inputs, outputs, and the potential environmental impacts of a product system throughout its life cycle." It assesses the life cycle based on greenhouse gases, acidification, eutrophication, resource depletion, primary energy, waste and toxicity (EUROPEN, ECR, 2009). An LCA can be used to make product life cycle maps and to identify improvement strategies (Fitzpatrick et al., 2012). The LCA is a useful tool since it covers the entire system and can be used to make well-supported design decisions, ones that lower the total environmental load. But the LCA has drawbacks because of its complexity and the difficulties of getting the proper input on so many factors in order to make fair and justifiable assessments (Pålsson et al., 2013). There are other, similar methods for evaluating the life cycle environmental consequences of a product and/or packaging from cradle-to-grave, such as the Resource and Environmental Profile Analysis (REPA) and Eco-Balance.

3.4.4.2 Sustainable packaging indicators and metrics framework by SPC

The Sustainable Packaging Coalition (SPC) has put forward a metrics framework to develop a set of common indicators and metrics to assess sustainable packaging as it is defined by SPC. The SPC definition holds criteria for material use, energy use, water use, material health, clean production and transport, cost and performance, community impact and worker impact (SPC, 2009). It is based on a life cycle perspective and includes to some extent the relation of the entire product and packaging system. But the integrated product and packaging view is only considered under the factor "cost and performance", where packaged product wastage is measured in terms of cost of lost product and also in the embodied energy from its production. While under the "material use" indicator, the focus is mainly on the direct packaging impact and not put in relation to the impact of product protection.

3.4.4.3 Comparative Packaging Assessment (COMPASS©)

This is an online software tool that allows packaging designers and engineers to compare the environmental aspects of up to four package designs. The software generates comparative profiles of the package options based on aspects such as fossil fuel consumption, water consumption, greenhouse gas emissions, and human and aquatic toxicity. The tool is best used as guidance or in benchmarks and does not calculate definitive answers, because it uses production-weighted, industry-average data in its calculations. The tool has its origins in the Sustainable Packaging Coalition (SPC) and was developed at the request of SPC member companies.

3.4.4.4 Packaging Impact Quick Evaluation Tool (PIQET©)

This is also an online software tool for packaging designers and engineers to rapidly assess the environmental impacts of different packaging options. The tool is based on life cycle assessment calculations using indicators including climate change, water use and solid waste and, a range of packaging specific metrics such as packaging to product ratio and recycled content.

3.4.5 Packaging design software and guidelines

There are several software solutions that in the hands of professionals are powerful tools in the packaging design process. Computer Aided Design (CAD), Computer Aided Manufacturing (CAM) and Computer Aided Engineering (CAE) are examples. In addition, some organizations provide specific guidelines for packaging designers and these can be used either as support in collaborations or as guiding principles for receiving certifications. Some of the software and guidelines that relate to packaging are mentioned briefly here.

3.4.5.1 Structural design and artwork software for packaging

There are several tools available, specifically designed for packaging professionals for structural design and virtual prototyping. Many of these are ideal tools for all kinds of corrugated, folding carton and different display designers. Many of these software programs are also excellent tools for achieving a more integrated and interactive view of the packaging artwork. Artwork professionals, however, often prefer to work with specific software programs that specialize in taking a design and turning it into a photo-realistic mock-up.

3.4.5.2 Packaging and palletization design software solutions

These software solutions help designers to evaluate and determine the product and packaging sizes and arrangement, pallet load and configuration, and load plans for transit vehicles like trucks, sea containers and rail cars. The audience for these software programs goes beyond packaging professionals and includes distribution and supply chain management professionals.

3.4.5.3 ECR packaging guidelines

Efficient Consumer Response (ECR) is a network of specialists in the fast-moving consumer goods industry who have come up with the FMCG packaging guide. It provides guidelines for the formulation and design of new packaging. The guide assembles and explains different standards and industry agreements, such as the International Standardization Organization (ISO) standards, the Swedish Standards Institute (SIS) module standards and the GS1 rules for information and labelling (ECR Sweden, 2012).

3.4.5.4 Organic and environmental labelling organizations

Most of these organizations provide guidelines and regulations for companies to use in order to be approved for using the certification labels. The majority of these guidelines and regulations focus on the product inside the package and not on the package itself. However, this is changing and initiatives for including packaging are under development. For example, in 2015, the Swedish organic label organization, KRAV, started to work on guidelines that include the package in the certification label. The guidelines are not yet published but they will include regulations on packaging materials, chemicals used and on the impact of packaging design on the reduction of food waste.

3.4.5.5 Global Protocol on Packaging Sustainability

The Global Protocol on Packaging Sustainability was developed by the Consumer Goods Forum (2011) to provide the consumer goods and packaging industries with a common language with which to discuss and assess the relative sustainability of packaging. The idea behind a common language is to facilitate businesses to maintain informed discussions and meaningful cross industrial cooperation.

3.4.6 Strategic guidance towards sustainable development

The above-mentioned tools and methods for insight mining, idea generation and for the evaluation and assessment of packaging are examples of what you may find in a toolbox for packaging design projects. All of these are of importance and each has its specific purpose. However, they all serve different tasks and steps in a design process, but none that really help the design team in navigating through the complexity of the contradicting demands put on packaging and the trade-offs. Nor do they consider the direct and indirect effects a package design has on sustainable development. Simply stated, there is no tool currently available that enables design teams to take a holistic approach to packaging design for sustainable development.

   To remedy this situation, a compass is provided in the next part of the book that is intended to assist packaging designers, managers and other decision-makers in making well thought through strategies and choices in packaging design for sustainable development. This is particularly the case when it comes to the next generation of packaging systems that need to be driven by innovation and sustainability, that should use resources "optimally", directly and indirectly, and at the same time fulfil the myriad of requirements and needs put on packaging. Packaging design is a balance of arts, crafts, science and technology, and design teams need to closely co-operate with stakeholders from other fields of expertise. Given that, the compass provided in the next part of the book will facilitate creativity and simplify decision-making between the different groups of experts.

References

Azzi A., Battini D., Persona A. and Sgarbossa F. (2012), Packaging design: General framework and research agenda. *Packaging Technology and Science*, 25, 435–456.

Ballou R.H., Gilbert S.M. and Mukherjee A, (2000), New managerial challenges from supply chain opportunities. *Industrial Marketing Management*, 29(1), 7–18.

Bramklev C. (2009), On a proposal for a generic package development process. *Packaging Technology and Science*, 22(3), 171–186.

Bruce, M. and Bessant J. (2002), *Design in Business. Strategic Innovation through Design*. Pearson Education, Harlow, UK.

Cachon G.P. (2003), Supply chain coordination with contracts. In: *Handbook in Operations Research and Management Science, Vol. 11: Supply Chain Management: Design, Coordination and Operation*, A.G. de Kok and S.C. Graves (eds), Elsevier, Netherlands, pp. 229–232.

Carlson S. (1991), *Executive Behaviour*. Reprinted with contributions by H. Mintzberg and R. Stewart, Studia Oeconomiae Negotiorum, Uppsala, Sweden.

Chiva R. and Alegre J. (2009), Investment in design and firm performance: The mediating role of design management. *Journal of Product Innovation Management*, 26, 424–440.

Colwill J., Wright E. and Rahimifard S. (2012), A holistic approach to design support for bio-polymer based packaging. *Journal of Polymers and the Environment*, 20(4), 1112–1123.

Consumer Goods Forum (2011), Global Protocol on Packaging Sustainability 2.0: http://www.theconsumergoodsforum.com/download-global-protocol-on-packaging-sustainability-gpps

Cross N. (2008), *Engineering Design Methods – Strategies for Product Design*. John Wiley & Sons, Chichester, UK.

ECR/Europen (2009), *Packaging in the Sustainabaility agenda: A Guide for Corporate Descision Makers*. ECR Europe and The European Organization for Packaging and the Environment (EUROPEN)

ECR Sweden (2012), Packaging Guide: http://www.ecr.se/lib/get/file.php?id=1531077f452692

Fitzpatrick L., Verghese K. and Lewis H. (2012), *Developing the Strategy*. In: *Packaging for Sustainability*, Verghese K., Lewis H. and Fitzpatrick L. (eds), Springer, London.

García-Arca J. and Prado J.C. (2008), Packaging design model from a supply chain approach. *Supply Chain Management: An International Journal*, 13(5), 375–380.

Grunert K.G., Jensen B., Sonne A.-M., Brunsø K., Scholderer J. et al. (2010), Consumer-oriented innovation in the food and personal care products sectors: Understanding consumers and using their insights in the innovation process. In: *Consumer-Driven Innovation in Food and Personal Care Products*, Woodhead Publishing Series in Food Science, Technology and Nutrition, pp. 3–24.

Harckham A.W. (1989), *Packaging Strategy Meeting the Challenge of Changing Times*. Technomic Publishing Inc., Pennsylvania, US.

Harry M.J. and Schroeder R.R. (2005), *Six Sigma: The Breakthrough Management Strategy Revolutionizing the World's Top Corporations*. Broadway Business.

Heskett J. (2002), *Toothpicks & Logos. Design in Everyday Life*. Oxford University Press, New York.

Hellström D. and Saghir M. (2007), Packaging and logistics interactions in retail supply chains. *Packaging Technology and Science*, 20,–216.

Hellström D., Johnsson C. and Norrman A. (2011), Risk and gain sharing challenges in inter-organisational implementation of RFID technology. *International Journal of Procurement Management*, 4(5), 513–534.

IDEO (2014), *Human-Centered Design – an Introduction*, 2nd Edition: www.ideo.org

ISO 14040 (2006) Environmental Management – Life Cycle Assessment – Principles and Framework: www.iso.org/iso/iso_catalogue

Jahre M. and Hatteland C.J. (2004), Packages and physical distribution: Implications for integration and standardisation. *International Journal of Physical Distribution & Logistics Management*, 34(2), 123–139.

Johnsson M. (1998), *Packaging logistics – a Value-Added Approach*. Doctoral thesis, Department of Engineering Logistics, Lund University, Sweden.

Jönson G. (1993), *Corrugated Board Packaging*, 1st Edition. Pira International, Surrey, UK.

Jüttner U., Peck H. and Christopher M. (2003), Supply chain risk management: Outlining and agenda for future research. *International Journal of Logistics: Research and Applications*, 6(4), 197–210.

Kano N., Seraku N., Takahashi F. and Tsuji S. (1984), Attractive quality and must-be quality. *Hinshitsu: the Journal of the Japanese Society for Quality Control*, 14(2), 39–48.

Lambert D.M. and Cooper M.C. (2000), Issues in supply chain management. *Industrial Marketing Management*, 29(1), 65–84.

Lambert D.M., Stock J.R. and Ellram L. (1998), *Fundamentals of Logistics Management*, International Edition. McGraw-Hill Higher Education, London.

Lee H.L. (2004), The triple-A supply chain. *Harvard Business Review*, 10(11), 102–112.

Lutters E., van Houten F.J.A.M., Bernard A., Mermoz E. and Schutte C.S.L. (2014), Tools and techniques for product design. *CIRP Annals – Manufacturing Technology*, 63(2), 607–630.

Martínez-Lorente A.R., Dewhurst F. and Dale B.G. (1998), Total quality management: Origins and evolution of the term. *The TQM Magazine*, 10(5), 378–386.

Mentzer J.T., DeWitt W., Keebler J.S., Min S., Nix N.W. et al. (2001), Defining supply chain management. *Journal of Business Logistics*, 22(2), 1–25.

Miaskiewicz T. and Kozar K.A. (2011), Personas and user-centered design: How can personas benefit product design processes? *Design Studies*, 32(5), 417–430.

Mintzberg H. (1973), *The Nature of Managerial Work*. Harper & Row, New York.

Mizuno S. (ed.) (1994), *QFD: The Customer-Driven Approach to Quality Planning and Deployment*. Asian Productivity Organization.

Molina-Besch K. and Pålsson H. (2016), A supply chain perspective on green packaging development-theory versus practice. *Packaging Technology and Science*, 29(1), 45–63.

Narayanan V.G. and Raman A. (2004) Aligning incentives in supply chains. *Harward Business Review*, Nov, 94–103.

Olsmats C. and Dominic C. (2003), Packaging scorecard – a packaging performance evaluation method. *Packaging Technology and Science*, 16, 9–14.

Osterwalder A. (2004), *The Business Model Ontology – a Proposition in a Design Science Approach*, PhD Dissertation, University of Lausanne, Switzerland.

Osterwalder A. and Pignuer Y. (2010), *Business Model Generation*. John Wiley & Sons, Hoboken, NJ.

Osterwalder A., Pigneur Y., Bernarda G. and Smith A. (2014), *Value Proposition Design*. John Wiley & Sons, Hoboken, NJ.

Plattner H., Mainel C. and Leifer L. (2011), *Design Thinking – Understand, Improve, Apply*. Springer, Heidelberg.

Plattner H., Meinel C. and Leifer L. (2015) *Design Thinking Research – Building Innovators*, Springer, Switzerland.

Pålsson H. and Hellström D. (2016), Packaging logistics in supply chain practice – current state, trade-offs and improvement potential. *International Journal of Logistics Research and Applications*: http://dx.doi: org/10.1080/13675567.2015.111547 accessed 19 June 2016.

Pålsson H., Finnsgård C. and Wänström C. (2013), Selection of packaging systems in supply chains from a sustainability perspective: The case of Volvo. *Packaging Technology and Science*, 26(5), 289–310.

Ravasi D. and Stigliani I. (2012), Product design: A review and research agenda for management studies. *International Journal of Management Reviews*, 14, 464–488.

Reid S., de Brentani U. and Kleinschmidt E. (2014), Divergent thinking and market visioning competence: An early front-end radical innovation success typology. *Industrial Marketing Management*, 43(8), 1351–1361.

Sustainable Packaging Coalition SPC (2009), *Sustainable Packaging Indicators and Metrics Framework, Version 1.0*. Greenblue, Charlottenville, VA.

Thomke S.H. and Von Hippel E. (2002), Customers as innovators. *Harvard Business Review*, 80, 74–81.

Twede D. (1992), The process of logistical packaging innovation. *Journal of Business Logistics*, 40(4), 85–88.

Twede D., Clarke R.H. and Tait J.A. (2000), Packaging postponement: A global packaging strategy. *Packaging Technology and Science*, 13(3), 105–115.

Ulrich K.T. and Eppinger S.D. (2007), *Product Design and Development*. McGraw-Hill Companies, Inc., Boston, MA.

usability.gov (2016); *Personas*. US Department of Health & Human Services: http://www.usability.gov/how-to-and-tools/methods/personas.html

Vazquez D., Bruce M. and Studd R. (2003), A case study exploring the packaging design management process within a UK food retailer. *British Food Journal*, 105(9), 602–617.

von Hippel E. and Katz R. (2002), Shifting innovation to users via toolkits. *Management Science*, 48, 821–833.

Witell L. and Löfgren M. (2005), Classification of quality attributes. *Managing Service Quality*. 17(1), 54–73.

Part II

A Packaging Design Compass for Sustainable Development

From Part I of the book we can draw the conclusion that packaging is an important part of our modern society, businesses and daily lives. The future role of packaging will almost certainly grow rather than decrease. Packaging is, and for a long time ahead will be, an essential field that will increasingly attract considerable interest from governments, institutions, companies and society on the whole.

Packaging is recognized as being a multidisciplinary academic field, and because of that, many aspects have to be considered in its design process. In practice, a majority of the professionals working with packaging or making packaging decisions do not consider themselves to be packaging professionals since they are affiliated with a specific organizational department, such as product development, production, logistics or marketing. These professionals often have different views, sometimes even conflicting opinions, on the packaging, depending on what departments they represent. As a result, they rarely understand the implications or unintended consequences their packaging decisions have on a bigger scale, such as that of sustainable development.

There are severe misconceptions about packaging from a sustainable development perspective. One misconception is that packaging is wasteful, not useful. Another is that overpackaging has severe consequences for sustainable development. The misconceptions result in conflicting attitudes towards packaging. Understanding these contradictions is not straightforward and often requires some kind of cognitive dissonance. However, for you to be able to consider sustainable development through packaging design, you need to accept: 1) that packaging is an enabler of sustainable development; and 2) the difference between "sustainable packaging design" and "packaging design for sustainable development".

Professionals who seek guidance on packaging design for sustainable development must first appreciate the complexity of packaging design, consisting of the many layers of aspects and requirements to consider that often go far beyond the package itself. In its nature, successful packaging design involves a holistic approach and a great deal of collaboration, compromising and respect about the needs and requirements of a variety of stakeholders, organizations and

Managing Packaging Design for Sustainable Development: A Compass for Strategic Directions, First Edition. Daniel Hellström and Annika Olsson.

distribution channels. Now that you have this knowledge and understanding, you are ready to be introduced to our packaging design compass for sustainable development.

Part II of this book – *A Packaging Design Compass for Sustainable Development* – is made up of two chapters. Chapter 4, "Introducing the compass" introduces you to the packaging design compass for sustainable development. It describes what the compass consists of, how to use it to navigate and orient yourself in the packaging design landscape, who is supposed to use it, and the research process in which it was created. In Chapter 5, "The directions of the compass", the different directions of the compass are described along with the potential impact the directions can have on sustainable development. The descriptions act as an introduction to Part III, where illustrative cases of all the compass directions are provided. They also offer you an opportunity to gain an in-depth view of the which of the compass directions can be useful for you.

4 Introducing the compass

It is necessary and obvious that we need to move towards a sustainable society and that contributions to this development need to be made on all levels of society. We strongly argue, in line with the current research, that it is technically possible and economically doable to make the transition to a more sustainable society through packaging design initiatives. What is needed is the right compass to guide us there. That is what we present to you here: a compass that irrespective of the type of packaging design project will guide you through the complexity of packaging in a world of fast and continuous change.

With our compass we encourage you to go off-road, to develop and innovate and remake the packaging solution that previously was best practice. Technology, people and organizations keep changing the routes we take to attain sustainable development. The compass is essential when you do not know which direction to take in a continuously and fast changing world. This is especially so in the field of packaging design, where some roads taken may contradict rather than contribute to sustainable development.

You might argue that a map could be used instead. But maps tend to be static and often become obsolete or constantly need to be redrawn. The road that leads you to success on one occasion, will probably not lead you there the second time around. As a result, the compass will help you find which direction to take.

4.1 Points of the compass

In contrast to an ordinary magnetic compass with its four directions (north, south, east and west), this compass has six directions in which you can pursue sustainable development through packaging design. The six directions are: protection; material use; fill rate; apportionment; user-friendliness; information and communication. They show the ability of packaging to achieve sustainable development through its design and are illustrated in Figure 4.1.

Just as in a magnetic compass, the directions are actually very interrelated. However, the needle on the packaging design compass does not point in opposite directions, like north and south on the traditional compass. This means that you can move in all directions at the same time. Within each packaging design

Managing Packaging Design for Sustainable Development: A Compass for Strategic Directions,
First Edition. Daniel Hellström and Annika Olsson.
© 2017 John Wiley & Sons, Ltd. Published 2017 by John Wiley & Sons, Ltd.

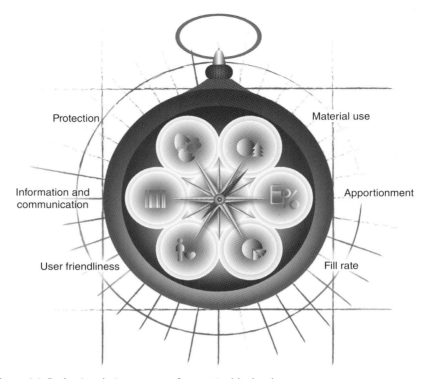

Figure 4.1 Packaging design compass for sustainable development.

direction, opposite outcomes can be assessed to determine if they have a positive or negative impact on sustainability. In practice, this means that when you move from one packaging design to another, all the directions will have changed more or less, affecting sustainability in a positive or negative manner. Thus, the compass does not simply point you in one direction, but in multiple directions. In fact, this is the whole point of using the compass. Its use is a balancing act, just as the needle on an ordinary compass needs to be balanced. In the best of worlds, you move in all directions in order to attain the best possible packaging solution for sustainable development. How to navigate and who directs the route when using the compass are explained in the rest of the chapter.

4.2 Users of the compass

In decision-making, we often navigate the landscape by instinct and fail to consider either the implications or unintended consequences. For professionals in the realm of packaging design, their knowledge and experiences often provide them with good instinct, which reduces their need for a compass. The same applies to companies with packaging as a core business. However, they face the risk of reproducing the same or similar solutions over and over, if they do not question their own established solutions. The packaging design compass can inspire the experienced professionals to go off-road and come up with more packaging solutions supporting sustainable development.

But, there is another important user audience that the packaging design compass addresses: professionals and companies that do not have packaging as a core competence or business. They can be suppliers or customers of packaging companies, such as product producers, retailers and all other companies that are affected by the design of packaging. Packaging design is a team effort in which people from multiple disciplines cooperative. Thus, there are many professionals from various disciplines, company functions and departments who are involved in packaging design decisions. They are professionals from R&D, production, marketing, sales, finance, purchasing, logistics and regulatory. For a majority of them, packaging is not their core competence or business. They need a compass to navigate the packaging landscape. If we go beyond professionals and companies, and consider policy-makers, interested organizations and society as a whole, a packaging design compass pointing towards sustainable development is necessary without a doubt.

4.3 How to navigate

In the very simplest of forms, the packaging design compass provides an overview of the directions that can potentially result in more sustainable solutions. Depending on the nature of the packaging design project and its context, you need something that points you in the right compass directions. As a decision-maker in different organizations and professions, you influence and determine what directions to take and what directions to prioritize. This decision is often based on your directives and the knowledge you have concerning strategy, technology, people, organizations, industry practices, businesses, supply chains and society. Your everyday activities, such as exchanging information, reviewing data, evaluating alternatives, implementing directives and following up, increase your knowledge and enable you to better decide what directions to take.

In the design process, decision-makers need to move their solutions in all directions of the compass and assess the impact on the three pillars of sustainable development to achieve the best possible packaging design. The packaging design compass is meant to be used as support for "multi-focused" decision-makers who put their energy into adapting the packaging design to its context and can use the compass to navigate in multiple directions. Thus, the directions are not to be considered as discrete options, nor are they meant to be used by "single-focused" decision-makers who strongly believe in taking only one course of action.

Equipped with the compass, you as the decision-maker can better navigate towards sustainable development in the packaging design landscape. The compass can be used on any packaging design project, from designing a brand-new solution to adjusting an existing packaging system in a line extension. It can also be used to implement sustainable development as a strategic directive for companies via the package design.

As described, the complexity of packaging design involves numerous aspects and requirements that need to be considered. Several of them often go beyond the packaging itself. That is one reason why packaging design is the product of teamwork and the different professionals involved. Each team member brings his or her knowledge and experience to the project. The project leader or manager needs to facilitate the team to share and merge their individual areas of expertise. In this setting, the compass is able to gather the team and provides clear, identifiable packaging design directions for the team to jointly and adequately address sustainable development.

4.4   The making of the compass – our methodology

Numerous years of conducting applied and multidisciplinary research in packaging logistics have contributed to the packaging design compass for sustainable development. A salient feature of the research process was the matching and combining of case study evidence and scientific literature in the fields of packaging, sustainability, design, logistics and supply chain management.

Our overall research methodology is best described as three distinct stages: the clarification stage, the descriptive stage, and the prescriptive stage. These three stages are heavily inspired by Blessing and Chakrabarti (2009) and are illustrated in Figure 4.2. The arrows between the stages illustrate the links and the many iterations carried out. The outcome of the clarification stage is presented in Part I of this book. The descriptive stage in Part III, and the prescriptive stage here in Part II.

The clarification stage increased our understanding of the state-of-the-art research and assisted us in formulating the core challenges in the field. In the descriptive stage, we identified, described and analysed numerous practical packaging design related cases and their impact on the three pillars of sustainable development. Some of these cases are presented in Part III as illustrative

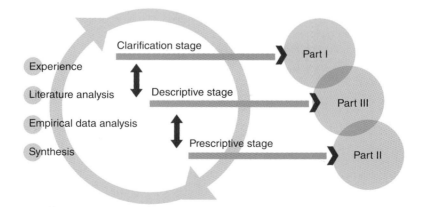

Figure 4.2 The research methodology that resulted in the packaging design compass for sustainable development, inspired by Blessing and Chakrabarti (2009).

examples of applying the six directions of the compass. In the prescriptive stage, we further analysed and discussed the cases, and developed the compass. Blessing and Chakrabarti (2009) pinpoint the core of this final stage by stating that it represents a "vision on how addressing one or more factors in the existing situation would lead to the realization of the desired, improved situation." Our vision took the shape of a packaging design compass for sustainable development.

When using the compass to navigate towards sustainable development there is a need to somehow assess the impact of packaging design solutions on the three pillars of sustainable development. This assessment can be done with different methods and tools, such as those presented in Chapters 3 and 4. The assessments in the illustrative cases are founded on analysing and identifying the positive or negative outcomes on the three pillars of sustainable development – planet, people and profit – along the supply chain or networks. The assessments are carried out in the descriptive stage and presented in Part III. Figure 4.3 presents the overall framework of the assessment features in the table format that is used in the cases. However, the framework was adapted to each case to adequately capture the impacts of packaging design on the three pillars of sustainable development.

From a scientific point of view, a salient feature of the research process was "abductive reasoning". This means that we matched and combined case study evidence and scientific literature in the fields of packaging, sustainability,

	Planet	People	Profit
Producer	+ Positive outcome	+ Positive outcome	+ Positive outcome
	– Negative outcome	– Negative outcome	– Negative outcome
Transport	+ Positive outcome	+ Positive outcome	+ Positive outcome
	– Negative outcome	– Negative outcome	– Negative outcome
Distribution Centre	+ Positive outcome	+ Positive outcome	+ Positive outcome
	– Negative outcome	– Negative outcome	– Negative outcome
Distribution	+ Positive outcome	+ Positive outcome	+ Positive outcome
	– Negative outcome	– Negative outcome	– Negative outcome
Retail	+ Positive outcome	+ Positive outcome	+ Positive outcome
	– Negative outcome	– Negative outcome	– Negative outcome
Consumer	+ Positive outcome	+ Positive outcome	+ Positive outcome
	– Negative outcome	– Negative outcome	– Negative outcome

Figure 4.3 Assessment framework used in the illustrative cases.

design, logistics and supply chain management. Abductive reasoning can be regarded as a search for theories from empirical observations, referred to as "systematic combining" or "theory matching" by Dubois and Gadde (2002) and Kovács and Spens (2005). This means that we went back and forth between our case studies and previous research in order to construct the compass.

References

Blessing L.T.M. and Chakrabarti A. (2009), *RM, a Design Research Methodology*. Springer, ISBN 978-1-84882-587-581.

Dubios A. and Gadde L-E. (2002), Systemic combining: An abductive approach to case research. *Journal of Business Research*, 55(7), 553–560.

Kovács G. and Spens K. (2005), Abductive reasoning in logistics research. *International Journal of Physical Distribution & Logistics Management*, 35(2), 132–144.

5 The directions of the compass

After introducing the compass and describing how you can use it in a packaging design project, it is time to present the directions in more detail. But before that, you need to keep in mind that a direction represents a packaging design capability for sustainable development. In other words, a direction defines an ability of packaging to contribute to sustainable development through its design. Even though each direction is described on its own, all six are strongly connected with each other. Irrespective of what type of packaging design project or what priority is placed on the three pillars of sustainable development, decision-makers must take all the directions into account. Not until then will you be able to consider both the implications and the unintended consequences of moving in any direction(s).

These directions are broad. Their impact on sustainable development changes from case to case, because each case is more or less unique. The directions are more than points on a compass. They can also serve as a source of inspiration to get you to start thinking in new ways and to start exploring the many opportunities found in the landscape of packaging design.

5.1 Protection

Unsuprisingly, protection is an important direction to take in order to secure product quality and safety throughout its life cycle, and to safeguard the surrounding environment from exposure to the content of the package. The most obvious reasons for protecting products are to avoid product waste and prevent damage. Regardless of how well-designed a product might be, damage during distribution or handling may cause it to be discarded before even being used. Avoiding such waste is imperative from a sustainable development point of view. For example, protection that decreases product waste is one of the most influential factors in reducing the negative impact that the production and distribution of goods has on the environment. Protection is thus a key direction of the packaging design compass, or rather a two-way direction. It is the protective interface between the product and the surrounding environment. The ability of packaging

Managing Packaging Design for Sustainable Development: A Compass for Strategic Directions, First Edition. Daniel Hellström and Annika Olsson. © 2017 John Wiley & Sons, Ltd. Published 2017 by John Wiley & Sons, Ltd.

design to protect a product and its surrounding environment can be devided into the following areas:

- to protect the product from mechanical stress – such as shock, vibrations or compressions;
- to protect the product from chemical and physical stress – such as sunlight, oxygen or humidity; and
- to protect the product from biological stress – such as microbiological spoilage.

For mechanical stress protection, the package needs to be designed so that it can withstand the stress that it will be exposed to in the package system on its way from production to the point of consumption. A key task for packaging designers is to know the circumstances in which the packed product will be distributed. It is evident that different contexts and different supply chains place different demands on packaging (Sohrabpour et al., 2016). Infrastructure issues such as road conditions and transport modes, and handling routines in different contexts, all affect the protection that is needed. Protection needs can also vary for the different levels of the packaging system. Packaging design for product protection is about the protection that the entire system offers. An increase in protection on the secondary packaging level, for example, can result in a decrease in the protection needed on the primary packaging level.

Protection is not only about protecting the product. It also protects the primary package so that it looks good when it arrives on the retail shelves. This is where the importance of designing adequately protective secondary and tertiary packaging comes in. Even if the product inside is intact, there are few consumers who want to purchase a product with a visibly damaged package. The protective role of packaging presents an opportunity for ensuring that products on retail shelves are sold and not wasted.

Physical, chemical or biological protection is first and foremost about constructing good protective barriers in primary packaging. This type of protection is mainly for edible products, but is important for other perishables such as pharmaceuticals, oils and chemicals. The issue for packaging designers here is to find the required barrier properties to create packages that maintain or better still, prolong the shelf life of the products. The protective role is to hinder any kind of exchange or transformation of substances between the product, the packaging material and the surrounding environment. The task is to protect the product so that it ends up in the hands of the consumer in a safe and hygienic condition, and so that it retains its quality throughout its shelf life.

5.2 Material use

Material use is an important direction of the compass because the selection of material in packaging design highly impacts sustainable development.

That we carefully need to consider material use when designing packaging systems for a more sustainable society is nothing new. The use of different packaging materials has received much attention from the industries that produce packaging, and from the authorities and consumer organizations. Material is needed in all packages and the selection of material in packaging design has a clear, direct impact on both the planet and profit pillars of sustainable development. This has made packaging material one of the most common ways of dealing with sustainability, but far too often with a suboptimized view of reducing the material. Inititatives that have been implemented to minimize packaging material are seen in, for example, shipping without packaging, reusable packaging systems, and selling products in bulk without packaging. However, from a packaging design point of view there are many other interrelated aspects to be considered to reach a solution that contributes to the three pillars of sustainable development.

The first thing to realize is that there is no superior or universal packaging material, since it is highly dependent on the requirements of the packed product and on the context and environment in which it is handled (Krochta, 2007; Robertson, 2013). When selecting packaging material in sustainable ways, it has to be done with care to ensure that it is efficiently used, while at the same time providing sufficient functionality to fulfill its tasks with the lowest required environmental impact for the entire system (Verghese and Lewis, 2007).

Due to the decreasing supply of global resources, the packaging industry is constantly working to improve packaging materials by making them lighter and by using more renewable resources. But lighter materials can never compromise the role of the packaging to protect and safeguard product quality. Material selection is a balancing act between finding the proper type and amount of material to achieve sufficient package functionality (such as product protection), and ensuring that the total resource utilization in the entire system does not increase.

As with all the other compass directions, material use has to be regarded from the entire packaging system point of view, meaning the primary, secondary and tertiary levels of the packaging system. For example, the intentions to develop packaging material with better protective properties needs to be balanced in relation to the protective properties in all levels. Adding material or selecting a stronger material on one level can result in less material being needed on another. The goal for all levels must be to use as few resources as possible in the entire system, and to select renewable materials as much as possible without sacrificing the functionality required for the packaging.

5.3   Fill rate

Fill rate is a strong packaging design capability for sustainable development. The most obvious potential impacts are less transportation, less damage, less and easier handling, less physical space needed (effective land utilization and energy consumption) across distribution channels, from raw materials to end consumers. All of this can contribute to sustainable development from the planet, people and profit perspectives.

Keep in mind that the fill rate influences the other compass directions in many ways that you need to consider. The following definition and explanation are provided to help you understand fill rate from a packaging point of view.

Fill rate refers to volume and weight efficiency. The fill rate is a typical packing problem. The aim is to find an arrangement in which the system is filled to as large an extent as possible. The proportion of space filled by this arrangement is called the fill rate. The most common challenge is to maximize the fill rate in the entire system. In other words, the fill rate is measured in how well a particular space is used. In order to take full advantage of the space available, the following measures need to be considered:

- The product should fill the primary package as much as possible.
- The relation between the primary packaging's outer volume and the volume of the product should be little as possible (sometimes described as inner fill rate).
- The primary packages should fill the secondary packages as much as possible.
- The relation between the secondary packaging's outer volume and the volume of the arrangement of primary packaging should be as little as possible.
- The secondary packages should fill the tertiary packages as much as possible.
- The relation between the tertiary packaging's outer volume and the volume of the arrangement of secondary packaging should be as little as possible (sometimes described as outer fill rate).
- The tertiary packages should fill the transport unit(s) space as much as possible.
- The tertiary packages should fill the warehouse location space as much as possible. A location is a physical space in a warehouse where an item of inventory can be held, such as a shelf or a bin.
- The secondary and/or primary packaging should fill the retail shelf (freezer, etc.) space as much as possible.
- The primary package should fill the consumer's shelf (refrigerator, etc.) space as much as possible.

For high-density products, it can be the weight rather than the volume that sets the limitation. Improving the fill rate is then difficult. However, if the product and the packaging system itself can be made lighter, this can result in potential fill rate improvements.

Stackability and stability are packaging abilities closely related to fill rate. If the packaging components can be stacked on top of each other, there are more possibilities to attain high fill rates in the packaging system, the transport unit, and so on. A high fill rate often increases the stability of the packaging system because the products and packaging components create a supporting frame which improves the ability to withstand potential forces of impact.

5.4 Apportionment

Apportionment is central in designing packaging systems for a more sustainable society. To apportion means to divide into parts according to a rule, typically based on the assumption that packaging design is a matter of quantity or volume. Figure 5.1 illustrates the number of items in primary packaging, the quantity of primary packaging in secondary packaging, and the quantity of primary or secondary packaging layered on a pallet or on a unit load carrier. Hence, apportionment

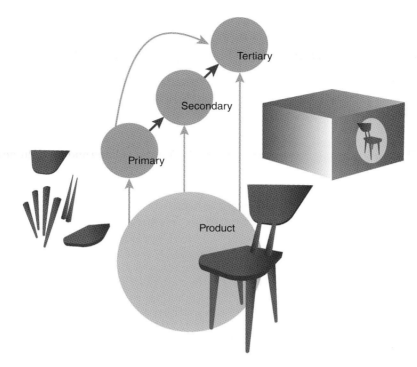

Figure 5.1 Typical apportionment decisions in packaging design.

is a rather straightforward direction in the packaging design compass, with the fundamental aim of providing the right quantity or volume. An intriguing question still remains: What is then the "right" quantity or volume?

An answer to that question would be that it depends on the product characteristics, such as value, size and weight. Another answer would be that the "right" quantity is in the eyes of the beholder. Different professionals and organizations show a diverse spectrum of apportionment. The different views need to be brought together, in one way or another, to support the three pillars of sustainable development. If not, something that seems so straightforward as apportionment will simply hinder sustainable development. In order to describe apportionment in an overall and integrative manner, we start out from two general perspectives: the supply perspective, and the demand perspective.

From a supply perspective, managers in production, operations, logistics or purchasing would argue that "what is right" about quantity goes beyond simply ensuring that the number of items in a box matches the packing slip; it has a major impact on costs and often affects the company's profitability and competitiveness. This is because fixed costs are incurred each time a product is put into production, ordered, handled or transported. The relative share of the cost decreases when the quantity is greater. On the other hand, the economics of scale is not always a good solution because large quantities cannot be stored indefinitely: The cost of inventory is also an essential part of the total cost. These managers often try to determine the optimal quantity of items to order in terms of minimizing the costs associated with the purchase, delivery and storage of a product. To do this they optimize inventory levels, work towards lead time reductions, forecast demand and take other actions that help them in ordering what they actually need: not more, not less. However, there are many other aspects that influence the "right" quantity, such as product variations, uncertainty in demand and lead time, and resource capacity. The major risk associated with quantity is obsolescence, which occurs when the product is no longer in demand or is unusable. This risk needs to be carefully considered since it results in discarded products, which negatively impact the planet and profit pillars of sustainable development.

From a demand perspective, the end consumers' viewpoint on the "right" quantity has additional apportionment impacts on sustainability. This is not an easy task, since different consumers have diverse consumption and situational-specific needs. Most often they are unable to clearly express or realize the needs they have. Yet, the amount of the product and the way it is placed in each sales unit is of significant importance. When it comes to medicine, for example, packaging can help patients (or through their caregivers) take the right dose of medication prescribed. At the same time, there are regulatory and safety aspects to consider about the quantity of certain medicines in a consumer package. In other categories, such as food, it is sometimes convenient for consumers to buy either small or big packs. Generally, smaller packages have more packaging material per packaged product, which considered in isolation is less environmentally friendly. Consumers appreciate a good price, which often persuades them to purchase the big pack rather than the separate smaller packs. From a sales perspective, this means an increase in the sales volumes of the products. But this can also have a negative environmental impact due to product waste if the consumer does not use or consume the contents before the expiration date. This is especially the case for perishable products that have a short shelf life or products with limited life cycles. The apportionment challenge is getting the product and package quantity to fit the variety of consumption needs of the targeted segments and for different consumption situations.

5.5 User-friendliness

The user-friendliness compass direction highlights the human interfaces with packaging and its contents. These interfaces go beyond the obvious physical interactions (ergonomics), and include perceptions and experiences in the use of packaging. User-friendliness deals with questions such as: How easy is it for users to learn and understand a task when they interact with a packaging design? How well can users perform a task? How correct is their use? How satisfying is it to use?

User and packaging interactions are often inadequate, whether in our homes as consumers or in our professions as retailers, distributors or producers. This lack of user-friendliness results in hurdles and problems. These can lead to unnecessary product waste, injuries, dissatisfaction and cost, to mention some of the consequences. Even the encouraging concepts that capture the user-friendliness direction, such as "Shelf-Ready" packaging, leave much room for improvement. The fact that Amazon has launched certified "Frustration-Free" packaging speaks for itself.

It is of great importance to acknowledge all the potential users when designing packaging. People very often only think of the end consumer when they speak about users, and forget about all the others. Packages are commonly handled by many different users along supply chains for different purposes and in different ways. All of them must be considered as users. No user is more important than any of the others.

Here is some general advice on how to improve user-friendliness:

- avoid double packaging;
- design packaging that can be opened in many ways without the use of dangerous tools;
- eliminate or reduce the strength requirements (such as carrying heavy loads with slippery surface);
- eliminate or reduce the need for fine motor skills (the use of small flaps and pull rings combined with gloves);
- be aware that users have far different preconditions – behaviorally (are used to handling certain types of packaging in certain ways) and physically.

The idea of practicing universal design, in the sense of aiming at designing "for all", is a step in the user-friendliness direction.

An increase in user-friendliness can positively affect several aspects of the packaging design output. There is no question that it contributes to the people pillar of sustainability. From a planet perspective, it contributes to the ease for users to reuse, recycle and dispose of packaging in the right ways. In addition, design that makes it understandable and easy for users to close, reseal, dispense and empty the packaging often saves resources by decreasing product waste. User-friendliness can lead to increased user efficiency (less time needed to accomplish a particular task) and satisfaction, and thereby increase productivity and sales and reduce support costs. This shows that it also has an impact on the profit pillar of sustainable development.

Once more, we assert that all the compass directions are interrelated. For example, in relation to material use, user-friendliness implies the use of packaging material that does not easily breakdown when opened, or the use of different materials and structures to achieve a good grip. Another direction that is closely related to user-friendliness and that considers the human interface with packaging as well, is the information and communication direction. Colours and symbols can inform and communicate to the user and help him or her understand how to safely open or carry a package.

5.6 Information and communication

In today's modern society we have an almost instantaneous flow and exchange of information and communication on a global scale. These flows impact all parts of society, beyond production and consumption, indicating perhaps that we are moving further away from the industrial era. The rise of the information age assigns new needs and requirements to packaging systems as well, and places new duties on the people involved in developing and designing packaging. With the fast development of information technologies and the global digitalization of information, the information and communication abilities of packaging are particularly important to consider in its design process.

5.6.1 Information abilities

There is no doubt that information on packaging can contribute to sustainable development. To start with, a firm's social, environmental and financial performance depends on achieving effective coordination and collaboration among people and organizations. These processes are extremely dependent on timely, accurate and transparent information. Information on primary, secondary and tertiary packaging is essential to getting things right from the beginning. Having some kind of automatic identification technology like barcodes or QR codes is a prerequisite today, and serves as a basis for organizations to access information about products, parcels, deliveries, shipments, etc., and share this information with others. These essential technologies are often directly printed onto the package or applied as labels. They play a key role in enabling organizations to automatically identify, track and locate products as they move through supply chains. These technologies encode standardized information such as product numbers, serial numbers and batch numbers. They not only enable easy access to far more information than a package has room for, but also the use of diagnostic and sensor indicators, such as time temperature indicators and freshness quality indicators.

It is hard to imagine what the world would look like without information on packaging. From an organizational point of view, packaging is an interface between the material flow and the information flow. The information on a package tells us things about its content, its past and its intended future; about its origin, where it was manufactured or assembled, who has shipped, sold, purchased or handled it and its destination. Such information is fundamental in managing complex deliveries that involve different modes of transport, consignment control (from container to pallet to individual box), knowing where shipments are and when and where they will arrive, and enabling the delivery of multiple shipments from various sources in a single drop.

From a user point of view, packaging is the physical interface between the product inside and the users who handle the packaged products along supply chains. The information on a package primarily tells us things about the content. This includes the kind of product, its expected shelf life, and its components or ingredients. The package information helps users by offering them instructions and advice on how to use the product, how the package is best handled during its use, and how packaging and the product should be disposed. All of this can contribute to sustainable development from the people, planet and profit perspectives. Examples of the potential impacts precise package information has on the people perspective are easier and correct handling (decreases e-mail and telephone calls about the product). From a planet viewpoint, information can reduce product waste (by decreasing incorrect handling, storage and use of the product) and damages (by decreasing picking errors and incorrect shipments that in turn reduce product returns), and result in reduced transportation (by decreasing the tracking of lost shipments). On the profit side, information on packaging influences the time to customer (by decreasing shipment delays) and sales (attractive product information). Even though there are many ways information can contribute to sustainable development, large amounts of information can be counter-productive and confusing – keep it short and concise.

5.6.2 Communication abilities

Packaging is a great medium for communicating with users and yet many designs still seem to ignore it, with the exception of graphic design. The communication abilities of packaging design can be divided into three categories: visual, verbal and tactile. When combined, all three have the power to sell products and build brands like nothing else. This combination can also contribute to the people and planet pillars of sustainable development.

Some of the communication elements of packaging design are size, shape, material, colours, typography, images, symbols, icons, layout and imagery. All of these affect user judgments and decisions, since they influence user perception of the product and the packaging. These design elements also aid in creating an image that has an impact on brand identification and the consumer mindset. Several of these elements can come together to create a visual impact. Colour contrasts, for example, have a great effect on text legibility and the comprehension of images and illustrations on packages. Photographs and illustrations are powerful elements that can be used to identify the differences in products and to communicate product functions, such as opening.

Since most packaging systems cannot talk, communication is primarily carried out through naming, descriptions and taglines that express a message or emotion. The tactile communication aspect is more of a physical user interaction, where material and sculptural forms convey product and packaging properties that in turn influence the users' perceptions of the quality of the image. In conclusion, this compass direction puts forward the silent salesman speaking to you – visually, verbally and physically – and is able to speak up about sustainable development as well.

References

Krochta J. (2007), *Food Packaging*. In: *Handbook of Food Engineering*, D.R. Heldman and D.B. Lund (eds), Taylor and Francis, New York.

Robertson G.L. (2013), *Food Packaging Principles and Practice*, 3rd Edition. CRC Press Taylor and Francis, Boca Raton, FL.

Sohrabpour V., Oghazi P. and Olsson A. (2016), An improved supplier driven packaging design and development method for supply chain efficiency. *Packaging Technology and Science*, 29(3), 161–173.

Verghese K. and Lewis H. (2007), Environmental innovation in industrial packaging: A supply chain approach. *International Journal of Production Research*, 45(18–19), 4381–4401.

Part III

Practical and Illustrative Cases

In the previous part of the book, you were introduced to the packaging design compass for sustainable development, where the different directions of the compass were described. We also explained how to use the compass when navigating and orienting in the packaging design landscape. The research process behind the formation of the compass was described as "systemic combining", where we went back and forth between a large number of empirical cases and linked them to research, literature and our valuable experiences in the field. It is now time to turn our attention to the cases that we have selected to illustrate the different directions of the compass.

The aim of presenting these illustrative cases is to provide you with insights and guidance to apply the compass in packaging design related projects. The cases are structured to challenge and inspire your packaging design thinking. All the cases exemplify and look at the different compass directions for a wide range of packaging systems. Individually, however, each case focuses on one specific direction of the packaging design compass in order to explain and provide a more in-depth perspective.

Three types of cases are presented to add additional perspectives to the compass directions: real-life, comparative and exploratory cases. The real-life cases describe implemented projects and their impact on the three pillars of sustainability. The comparative cases contrast two packaging systems and the potential sustainability impacts are discussed. The exploratory cases illustrate something innovative that is not yet implemented or on the market, so they are not found in practice.

The compass did not exist when the cases where developed, but has been applied in retrospect. But we would argue that an "inner" compass has been used by those who participated in the real-life packaging design projects or who were thinking about the new innovative solutions that would certainly contribute to sustainable development if put on the market. Whatever the case, the main rationale behind them is to inspire you to think about the different directions you can take in your own packaging design projects. We realize that it is quite a challenge to try to consider all the directions at the same time, since some of them may be contradictory, resulting in trade-offs between the different directions that need to be taken.

The cases presented deal with packaging systems for a variety of products and categories. The majority are concerned with product and packaging systems in the fast-moving consumer goods industry, especially in the food sector. There are some cases that consider durable goods, unit load carriers, electronics and non-consumer products. Food was chosen as it has a fundamental position and enjoys a central importance in our society because it so basically relates to health, happiness and political stability. The food sector is the largest economic sector in the European Union and represents large volumes of distributed goods around the globe. The biological nature of food products makes them sensitive and poses challenges when it comes to their handling, their limited storage time, and that they spoil easily. Food packages are also under particularly high cost pressure from product producers and retailers. At the same time, caution needs to be taken in relation to human health and safety, because food is something that we put into our bodies.

Part III of this book – *Practical and Illustrative Cases* – consists of six chapters. Each chapter is about a specific compass direction and presents three unique cases. In total, 18 cases are presented. All cases are standalone, so the reader can pick and choose the ones they find to be interesting and relevant. Keep in mind that each case highlights a specific compass direction, but often includes and considers others, since all of them are interrelated. Enjoy the cases!

6 Product protection

There is an extensive amount of research evidence highlighting that there are considerable sustainability implications related to the ability of packaging to protect products (for example, Bertoluci et al., 2013; Lindh et al., 2016; Wikström and Williams, 2010; Williams and Wikström, 2011). Even though it is one of the most central directions in theory, product protection for reduced product waste is not always considered in practice. In this chapter, we look into three cases that have the potential to lower product waste by means of well thought through protective packaging design. The cases describe the impact that increased product protection has on sustainable development. The first two cases involve fruit, a product type with generally high levels of waste. The third case is about a sofa from IKEA. It shows how IKEA deals with product quality and protection in relation to different packaging solutions.

The reason for highlighting fruit in two cases is that the fruit and vegetable industry is characterized by sensitive fresh produce, global sourcing, high levels of waste and a rather low product value that must be cost-effective. It is also so that the fruit and vegetable departments in grocery stores generate the most food waste (Eriksson and Stridh, 2011; Gunders, 2012).

The first comparative case is about packaging for grapes from South Africa. The case describes and compares the impact the package design has on the quality of the grapes sold in Sweden. It also highlights the impact that unsold grapes have on the farmers in South Africa. The second case is comparative, about citrus fruits that are also imported to Sweden from South Africa. It is a case that shows the role packaging plays in lowering waste with protective packaging. The third real-life case emphasizes that it is not only fresh produce in which packaging plays a key role in protecting the products. Product protection is an essensial compass direction for all kinds of products. This case illustrates how the global retailer IKEA works with product quality and packaging for their most sold sofa model, the "Ektorp". In high-volume furniture supply chains like this, product protection has a significant impact on whether the products are being picked on the shelves, as well as on return rates for product claims.

All three cases, whether fruits or sofas, clearly demonstrate that there are significant sustainability implications related to product protection via the packaging systems.

Managing Packaging Design for Sustainable Development: A Compass for Strategic Directions, First Edition. Daniel Hellström and Annika Olsson.
© 2017 John Wiley & Sons, Ltd. Published 2017 by John Wiley & Sons, Ltd.

The cases are:

- Better quality grapes for the people;
- Cheap is not always the best: The citrus box;
- IKEA *Ektorp* sofas: Knock-down boxing.

6.1 Better quality grapes for the people

Case by Fredrik Nilsson

Serving and eating fresh, sweet and seedless table grapes is popular in many homes. According to the Food and Agriculture Organization (FAO, 2002), 75,866 square kilometres of the world are dedicated to growing grapes. Approximately 71% of world grape production is used for wine, 27% as fresh fruit, and 2% as dried fruit. While there is an increase of wineries in the northern hemisphere, table grapes grow best in warmer climates with an average temperature at 15°C or higher. Consequently, in order to satisfy consumer demands in northern Europe, table grapes have to be imported.

In Sweden, the retailers import grapes from different countries at different times of the year, depending on where it is the harvest season (Table 6.1). The harvest season can change from year to year, from continent to continent, or based on weather conditions. The shelf life of the grapes depends on how they are treated and stored, but normally last 4–5 weeks from when they are harvested.

Table grapes are a sensitive and highly seasonal product that needs special consideration in the supply chains in order to reach the consumers intact. At the same time, grapes are an inexpensive consumer product with a great many suppliers to choose from. So, in order to meet consumer demands for table

Table 6.1 Countries from where Swedish retailers import grapes during different periods.

Period	Import countries
Dec–Feb	Namibia
	South Africa
	Argentina
Mar–May	Egypt
Jun–Oct	Greece
	Spain
	Italy
Nov	Brazil
	Peru

grapes and not lose sales opportunities, the focus has been on supply chain efficiency by the industry. But to mitigate the risks of products in the stores that are past their prime, high order quantities are the norm. For most of the importing companies and retailers this is not a problem, but as highlighted by Ras and Vermeulen (2008) in their study of South African producers of table grapes, "The South African producer bears the market risks of the produce until it is sold to the consumer in the European shop and then only is the money transferred to the producer." Most producers (66% in the study) are not paid for the table grapes that are damaged or go unsold in Europe. Consequently, all damage that occurs to table grapes downstream the supply chain is out of the producers' control, at the same time as they only receive payment for what is sold in retail outlets. With supply chains consisting of many different actors with operations that vary in quality, the impact on the producers is significant. As a result, product protection by packaging can have a great impact in reducing waste or damage to the product, directly resulting in better financial results for the producer.

This case starts with farmers in South Africa and follows the handling of grapes through one exporting company to the European importer and then to Swedish consumers via one wholesaler and one retail company.

6.1.1   The table grapes packaging system

The packaging system for table grapes consists of three levels of packaging: a consumer package (primary); a retail package (secondary); and a transport package (tertiary). One of the most common consumer packages is made of PET (polyethylene terephthalate) and carries 500 g of grapes (Figure 6.1). The transparent PET unit is $190 \times 115 \times 80$ mm and weighs approximately 24 g. The material is commonly used as a barrier for gases, moisture and solvents. PET is recyclable since it is a thermoplastic material which can be remelted.

Since the grapes mould easily (Christie, 2001), the consumer package needs to have some openings so that air and humidity can enter and exit the package. In the case of the PET boxes, this is solved by a number of circular holes made in the bottom of the box, as shown in Figure 6.1. In order to minimize mechanical damages, the package also needs to be rigid and be the right size so that the fruit cannot move around. Hence, the PET box functions as an apportioning and protecting unit that safeguards the grapes from mechanical damage and enables the necessary air flow.

The retail or secondary package consists of boxes made out of corrugated board. They hold 10 consumer packages in one layer or 20 in two layers and have a fully open top or are partially covered at the ends to enable better stacking. In some boxes, the consumer packages are stacked inside of a plastic cover and/or a thin paper is folded over them for protection. Inside the cardboard boxes a preservative sulphur dioxide (SO_2) insert is placed in order for the products to last longer (hampering the moulding process) during transport and

Figure 6.1 The PET, 500 g consumer package for table grapes, showing the circular holes for airflow.

storage. SO_2 is the fungicide treatment generally used in the grape industry (Christie, 2001).

When the grapes are imported from countries outside Europe, it is common that they arrive on non-standardized pallets as well as in different quantities of grapes. When the grapes are transported through Europe by truck to wholesalers in Sweden, the corrugated boxes are stacked on EUR1 pallets (800 × 1,200 mm) in 4 columns with 18 cardboard boxes on each. The total height is 1,800 mm and leaves room for the air to circulate in the container. The retail packages are stabilized with plastic wrapping. It is important that the grapes are not totally covered with the plastic wrapping, so that air can circulate and so that no condensation occurs from cooling.

6.1.2 The table grapes supply chain and challenges

The supply chain involving the distribution of table grapes from South Africa is similar to other supply chains of grapes. The supply chains differ in the type of shipment, depending upon where in the world the grapes come from. Table 6.2 provides an overview of the supply chain actors involved and the related logistics activities in the flow of grapes from South Africa to Sweden.

The fruit is harvested unripe in order to extend its shelf life. The grapes are harvested during the morning hours. since the temperature increases during the day and can damage the quality of the fruit. The grapes are directly transported to the fruit manufacturer where they are cooled down to a temperature of 15°C. All the direct handling of the fruit needs to be executed manually, since grapes are a sensitive product. The clusters are cut to the right size, and damaged and rotten fruit are removed. The fruit is then packed into primary, secondary and tertiary packaging and labelled for distribution and sales before the being moved to a cooler. The cooler maintains a temperature of 0–2°C, which is best for a sustainable shelf life. The packaging process should be performed rapidly in order to get the fruit into the right temperature as quickly as possible. The grapes

Table 6.2 Illustration of the supply chain actors from South Africa to Sweden and their activities for different packaging levels.

Supply chain actor	Logistical activities	Primary packaging	Secondary packaging	Tertiary packaging
Producer	Harvesting			
	Packing		×	×
	Transport			×
Fruit manufacturer	Receiving			×
	Storing		×	×
	Packing	×	×	
	Shipping			×
Fruit exporter	Receiving			×
	Storing		×	×
	Packing	×	×	
	Shipping			×
Fruit importer to Europe	Receiving			×
	Storing		×	×
	Packing		×	
	Shipping			×
Wholesaler	Transport		×	×
	Storing			×
	Picking		×	×
	Shipping			×
	Transport			×
Retailer	Receiving		×	×
	Replenishment		×	×
	Re-use		×	×
Consumer	Picking	×	×	
	Transport	×		

are then loaded into trucks or containers with a refrigeration unit that should keep the temperature at approximately 1°C. They are also required to be equipped with an air flow and humidity system that maintains a humidity of 90–95% in the truck or container.

The fruit producer in South Africa ships the table grapes by truck to the harbour from which they are transported by boat to Europe. From the European importers they are transferred by either boat or truck to wholesaler distribution centres in Sweden. Shipments from countries within Europe take four days and shipments overseas takes approximately three weeks. The major part of the lead time is during transports, especially longer ones by boat. The boat transports can be up to a week late due to bad weather or ineffective logistics in harbours.

The role of the packaging system for grapes is essential in transporting the product over the globe. Unfortunately, the different levels of the packaging systems are often treated separately, which means that the primary, secondary and tertiary levels of packaging are seldom optimized together. Taking a system view of the packages and the supply chain reveals challenges that influence both the effectiveness and efficiency of the flow, as well as the possibility to reduce product waste. For example, while the primary level is designed for airflow, the secondary and the tertiary levels (boxes and pallet with plastic wrapping) hamper this effect in most of the storage, handling and transport in the supply chain, since the overall packaging system is not designed to optimally handle the airflow. One of the global traders of fruit reports that they are having quality problems with the cardboard boxes for both 10 and 20 PET packages. The packaging systems for these boxes are seldom designed together and are often driven by different departments, such as marketing and logistics, where the former focuses on sellability and the later on efficient flow. By designing the packaging system together (as further elaborated in Part I of this book) a real improvement in airflow can be attained leading to less waste and more sold table grapes.

It is also found that since the holes are circular in the PET boxes, the grapes can completely block these holes and hamper the air flow that is central for minimizing mould production. Making these holes square-shaped instead would be a rather simple redesign to secure airflow and lower the waste due to mould problems.

A quality control is carried out when the grapes arrive at a distribution centre. The grapes that are approved are then stored in the area for fruit and vegetables which has a temperature between 2 and 4°C. The grapes that do not pass the quality check are registered and sent back or thrown away and the wholesaler/retailer is financially reimbursed (approximately 1.5% of the grapes do not pass the quality check at the wholesaler). Before shipping the grapes to retail outlets, a second quality control is carried out when picking the goods. The distribution centre ships either full pallets (often the same that the grapes arrived on) or container wagons to the retail stores depending on the order size. The grapes are mixed with other fruits and vegetables to maximize the space utilization of the container wagons. This is followed by transportation to the retail stores along with other products.

Depending on the size of the store and the demands of the retailer, the grapes are ordered and arrive at the store on full pallets with the plastic wrapping still on, or on grid wagons along with other products. A quality control is carried out upon arrival at each retail stores. One of the Swedish wholesaler distribution centres estimates that approximately €130,000 worth of grapes annually do not pass the quality check at the retailers. In larger retail stores the pallet goes directly out to the fruit area, the plastic wrapping is removed and the secondary packaging is taken off. The pallet can also be placed in cold storage in the retailer storage room. For small- and medium-sized retailers that usually receive retail packages on a mixed wagon, the grapes are usually taken straight out to the store. If not, then cold storage is used.

The consumer in Sweden can purchase 500 g of grapes for a price between €1.6 and €3.75, depending on which country the grapes come from and which type of store they are purchased in. The price is also dependent on the type of the grape, changes in climate, exchange rates, transport type, duties, etc. But most of all, it is supply and demand that controls the market price.

For the table grape farmers in South Africa, especially those with special contract agreements, the amount of produce sold impacts their incomes and their daily lives. For most of the actors in the supply chain, the waste in kilograms is not an issue as long as they can be economically compensated. Instead, the effect of such practices is that the wholesalers and retailers order more than necessary so that they can still make their margins and adequately satisfy the home market consumers with table grapes. At the same time, the farmers have to produce an excess of grapes for both consumption and waste, yet only get paid for the portion consumed.

6.1.3 Supply chain implications

The food industry has been polarized in terms of the power shift towards the retail chains. The consequences for the other supply chain actors have been, as one might imagine, both good and bad. In the case of table grapes, there are 450+ producers in South Africa that are all dependent on the efficiency and effectiveness of the supply chain as well as the packaging logistics solutions in order to benefit and survive. The export markets are necessary for survival and provide opportunities for the producers and farmers to diversify their production units and generate income. Yet they have no direct influence either on the supply chain or on the packaging suppliers. This is because many producers in South Africa are competing not only with one another, but with suppliers in other regions of the world as well. This means that the retailers can place high demands on the importers for quality assurances, among others things. The importers in turn forward these demands to the producers and at the same time play them off against each other in terms of price.

By focusing on the improvements in packaging that can increase protection and facilitate more efficient handling, the potential for fewer damaged products (which means less waste) is great, and both the economic and social aspects for the producers can be improved (Table 6.3). A major challenge for this to take place is in getting the importers and retailers involved and supportive of such developments. In addition, increased interaction between farmers, producers and the packaging suppliers can lead to improved solutions based on accurate supply and consumer demands. Today, the waste of table grapes (in non-financial terms) is not considered a "real problem", as it is handled in financial terms by the powerful actors in the supply chain: The returns and waste that are not sold are not paid for, and the cost of handling and quality assurance at the retailers is regarded as a minor aspect or is not even assessed. But from the environmental point of view, the waste of products that have been produced and transported around the globe is devastating.

Table 6.3 The impact (+ pros and − cons) on the three pillars of sustainability, on the suggested packaging design of table grapes.

	Planet	People	Profit
Producer	+ Less product waste	+ Improved economy for farmers	+ More products to the market
			− Increased packaging cost
Transport	+ Fewer damaged products	− Fewer truck drivers needed	+ Higher resource utilization
	+ Fewer transports		
Distribution Centre	+ Less warehouse space needed	+ Fewer problems with broken packages	+ Less packaging material to handle
		− Less personnel needed	+ More products per delivery
Distribution	+ Higher resource utilization		+ Higher resource utilization
Retail	+ Less product waste	+ Fewer damaged products to deal with	+ More products sold
Consumer	+ Less product waste	+ Higher quality	+ Less product waste

Case acknowledgements

Special thanks go to Tania Nieuwoudt, Malin Göransson, Samin Houshmand, Joakim Lewin and Lukasz Szychlinski for the initial work carried out in this case.

6.2   Cheap is not always the best: The citrus box

Case by Fredrik Nilsson

Oranges and other citrus fruits are popular in most markets and a must in retail stores. The citrus trees bear fruits of different shapes and sizes, which are full of flavour, juice and fragrance. They have been in existence for more than 300 years. Major commercial citrus growing areas include southern China, the Mediterranean Basin (including southern Spain), South Africa, Australia, the southernmost United States, Mexico and parts of South America. South Africa's citrus production is in focus in this case. The citrus industry in South Africa is mainly export oriented and there are more than 20 million citrus trees planted on 58,000 hectares of land. The citrus trees in South Africa produce fruit from March until October.

The normal shelf life for citrus fruit is one week. If they are refrigerated they stay fresh for about one month. Citrus fruit is sensitive to impact and pressure, which means that damage to the fruit is a big concern for the farmers. Like most other fruits, citruses can easily be damaged, making them impossible to sell. The fruit can be damaged in several ways. One results from the manner in which the fruit is packed. Consequently, protection is central in the supply chain by means of careful processes and protective packaging solutions.

This case evaluates the packaging system and supply chains of citrus fruits in South Africa up to the point the fruit is either loaded into a container for international transport or distributed to regional retail stores.

6.2.1 The citrus packaging system

The citrus fruits at the case company, Cape Citrus, are categorized into two divisions: hard and soft peel. During the packaging process hard peeled fruits are packed up to 6 layers, compared to the soft peeled fruits that can only be packed up to 3 layers. The packages consist of corrugated boxes that are stacked on pallets for storage and transportation. Different market demands are placed on the boxes to meet the specific requirements of the packaging of fruits.

This case focuses on boxes that carry 15 kg of citrus ($600 \times 400 \times 170$ mm). They were determined to have the most potential for improvement according to the results from the scorecards and interviews at the supply chain actors. These boxes were identified by the retailers as causing excessive waste and thereby needed to be improved.

6.2.2 The citrus supply chain

Cape Citrus, which is part of the Suncape Group, markets and packs citrus from three different producers located in the Western Cape in South Africa. Cape Citrus products are distributed all over South Africa, as well as to other countries and markets. Each of these markets has different specifications and requirements. The company has outsourced its packaging to a contract packer named Stellenpak. Stellenpak has packing, cooling and cold storage facilities and all transportation and shipping arrangements are made by the company's Export Administration Department. Stellenpak uses a logistics service provider (LSP) for all transportation needs. The LSP delivers to various retailers, including Woolworths. The final actor in this supply chain are Woolworths' customers. For the export markets, such as the Swedish market, there are a few more actors involving an importing company for the European market and then wholesalers that distribute the fruit to domestic markets.

6.2.3 Key handling activities

Three key activities are necessary in the handling of citrus fruit in order to provide high-quality products: 1) the packaging process; 2) re-cooling and dispatch; and 3) quality control.

6.2.3.1 The packaging process

The sorting of the citrus fruit is done both manually and automatically. The fruit is sorted by a system where weight, colour and size are used to sort the fruit into groups. The soft peel citrus is packed in smaller boxes with less fruit per box, with wrapping between the layers to separate the fruit. This minimizes the pressure by limiting the mechanical damage to the citrus. The hard peel citrus is placed in bigger boxes with more fruit (endures more pressure), also with wrapping between the layers used to separate the fruit.

Stellenpak packages the fruits from Cape Citrus in cardboard boxes, stacks them on a wooden pallet, wraps the pallet in plastic and then uses a third-party logistics provider to deliver it to the Woolworths' retail outlet. Woolworths either sells the fruit in the cardboard boxes or unpacks it so the customer can manually pack it into a plastic bag.

6.2.3.2   Re-cooling and dispatch

The re-cooling process is where the packed cardboard boxes are cooled to the optimum storage temperature between 7.0°C and 10.0°C. Accurate measurements, management, record-keeping of products, and storage conditions are some of the activities required in the cooling and dispatch procedure.

In the beginning of the season, citrus is stored at 10°C to enhance the colour development. The temperature should be 7.0°C when the citrus is shipped. Citrus in transit needs to be pre-cooled and shipped at 0.0°C, depending on the requirements of the importing country.

6.2.3.3   Quality control

Cape Citrus packs their fruit according to the Producer Unit Codes, where the information is captured on a real-time pallet tracking computer system and printed on the cardboard boxes, ensuring that the products are traceable and identifiable. Cape Citrus complies with South African and international specifications (GLOBALG.A.P., HACCP, PPECB[1]) in terms of the handling and storing of fresh fruit.

6.2.4   Challenges in the citrus supply chain

Damage to the citrus is a major concern for the farmers. When damaged, it cannot be sold. The packaging process plays a central role in this, as does handling in the supply chain. If the citrus is not carefully packed, it gets damaged and loses its value. The cost of the packaging is a significant factor in the price of the product, along with transport and storage. Consequently, a balance needs to be found between low cost and high protection.

A corrugated board box is currently used for the citrus fruits. That is an inexpensive way of packing the citrus fruit and thus keeps the total packaging cost low. A total of 65 boxes are stacked on a pallet. The height of the pallet is 2.4 m, which results in high cube utilization.

Based on a supply chain analysis and the use of packaging scorecards to collect packaging related data, a number of areas for improvement of the corrugated board box solution were found, all of which could lower product waste. These included:

- the strengthening of corners;
- different types of corrugated material, including the use of recycled material; and
- ensuring that cardboard waste or overpacking does not take place.

---

1 The Global Partnership for Good Agricultural Practices, Hazard Analysis Critical Control Point, The Perishable Products Export Control Board

In addition, two more initiatives are suggested for efficiency improvements in the supply chain:

1) time and transport measures to decrease the environmental impact;
2) the information flow between the different role players about the annual/seasonal harvest.

The main packaging design problem identified was the impact on the lower level boxes for the 2.4 m high stacked pallets. During transport and handling, the pressure from the high loads causes the lower three levels of boxes to bend or totally collapse (16–19 boxes of the 65 boxes are damaged), which in turn results in damaged citrus fruits that can no longer be sold to the proposed markets. This results in waste in the supply chain. Generally, this leaves the exporter with a loss of up to 30% of the original citrus fruits that were available to sell. Not only does the company lose money they would have received for the fruit, but also the money they have spent on the preparation of the fruits thus far, and the space that the damaged fruits took up on the original pallets. On the domestic market it means that the retailer is reimbursed for up to 30% of the products per delivery from the LSP or the producer.

6.2.5   Supply chain implications

The implications of this case are several (Table 6.4). The lowering of product waste with better protection of the package would positively impact the farmers and the food producers primarily. However, there are additional costs

Table 6.4 The impact (+ pros and – cons) on the three pillars of sustainability, from a reinforced cardboard box.

	Planet	People	Profit
Producer	+ Less product waste	+ Improved economy for farmers	+ More products to the market
			– Increased packaging cost
Transport	+ Fewer damaged products	– Fewer truck drivers needed	+ Higher resource utilization
	+ Fewer transports		
Distribution Centre	+ Less warehouse space needed	+ Fewer problems with broken packages	+ Less packaging material to handle
Distribution	+ Less product waste	+ Fewer broken packages to handle	
Retail	+ Less product waste	+ Fewer damaged products to deal with	+ More products sold
		+ Fewer unstable packages to deal with	
Consumer	+ Less product waste	+ Higher quality	+ Less product waste

that need to be covered because a stronger corrugated board or a new packaging solution will be more expensive.

But the reduced waste will outweigh the increased packaging cost. By either reinforcing the entire box with a stronger corrugated solution or by adding extra material or inserts in the corners for the lower levels, the packaging could improve the number of products that are successfully transported in the supply chain and potentially sold to the consumers. Currently, the loss of up to 30% of fine products due to cheap packaging solutions has financial, social and environmental consequences. For example, a loss of 16 boxes with 15 kg of oranges per box results economically in a loss of €387/pallet (based on a sales price of €1.61/kg), and environmentally in a loss of 60 kg CO_2e/pallet (0.25 kg CO_2e impact per kg oranges) (Wallén et al., 2004).

However, with the dominant focus on cost and efficiency in the food industry along with limited knowledge about packaging logistics, inexpensive packaging solutions are selected even though they cause large losses of products due to minimal protection. In order to develop sustainably this has to change because it is not only the financial aspects that should be emphasized, but also the actual product waste and the integrated view of products and packaging. The cheapest packaging material or solution is not always the best for the entire system!

Case acknowledgements

The author thanks Gert Mans, Wilmarie Giliomee, Carel Cilliers and Edgar Fourie for the suggested improvements and initial analysis in this case.

6.3 IKEA *Ektorp* sofas: Knock-down boxing

Case by Daniel Hellström

The global retailer IKEA is well known for its flat packages and for its efforts to improve logistics and supply chain performance by focusing on packaging design (Hellström and Nilsson, 2011; Jonsson and Kalling, 2007; Klevås, 2005; Klevås et al., 2006). One of IKEA's fundamental packaging requirements is to protect its products the entire way from the producers until they are assembled at home by the consumers. When a product or its packaging is damaged along the supply chain, it is not only the end consumer's experience that is drastically affected, it also requires additional work on the company's part and, ultimately, results in product waste.

In 2009 IKEA introduced a knock-down version of the icon and global bestselling sofa, *Ektorp*. It had previously been distributed and sold to consumers assembled. The knock-down version in a flat package has been a success, because it reduced the transport demand by approximately 400,000 m³ a year. However, after a year on the market, the costs resulting from customer complaints of poor quality were considerable. An assessment revealed that more than 50% of the poor-quality costs were directly related to packaging. Thus, IKEA initiated a packaging redesign project with the motivation to improve quality.

6.3.1 The previous packaging solution

The previous packaging solution for the knock-down version of *Ektorp* was based on the classic regular slotted container, a simple and low-cost solution. The container dimensions were $870 \times 400 \times 2020$ mm, sealed with transparent tape. Three of these were placed on a unit load and stabilized using stretch wrap and straps. Figure 6.2 shows what this packaging system solution looks like when it is handled carelessly. The solution had been developed from the mindset of minimizing the cost of packaging by finding the cheapest packaging solution available. For IKEA, it is very uncommon to use a regular slotted container this large as a primary package, especially for a product that is heavy, non-self-supporting and globally distributed. The quality problem with this packaging solution for *Ektorp* resulted in product damages throughout the supply chain, which in turn resulted in additional work for the employees and waste before the product even reached the customer.

6.3.2 Managing damages

All types of product-related damages are reported globally throughout IKEA's supply chain. Working procedures at the various stages of the supply chain have been put in place to prevent damages and to deal with those that occur in an

Figure 6.2 An example of damage to *Ektorp* resulting from poor packaging.

efficient manner. For example, the stores are not supposed to accept products that are already clearly damaged at the time of delivery. These products are to be reported and sent back. Unfortunately, this procedure is tedious and not always put into practice. This results in damaged products ending up in the stores. There are several ways to handle damaged products at IKEA stores. This depends on the product and the level of damage. Disposing of products is not the IKEA way. A majority of the damaged goods are reduced in price and sold in the bargain corner along with ex-display items and discontinued products.

Unsurprisingly, some of the damages are related to packaging. Damaged packaging implies that the product has been handled carelessly and roughly. End consumers pick products with "neat and clean" packaging compared to those that are dirty, opened, scratched, torn, squeezed, or where the tape is peeled off. Even when the product as such is in perfect condition, if the packaging is damaged, it will not sell. IKEA stores deal with this by repacking the product at stores using a packaging cutting table.

6.3.3   Packaging redesign and impacts

The improved packaging solution for *Ektorp* is based on a new structural design that has a bottom and top that form a box (Figure 6.3). This is a self-supporting unit, because the knock-down sofa itself cannot serve as a supporting element. Straps are used to seal the box and it has the same dimensions as the previous box. Before it was implemented, several field tests were carried out to validate that the new solution performed as intended. One year after the implementation of the improved solution it had resulted in drastically reducing damage at all stages in the supply chain.

Figure 6.3 The new structural box design for *Ektorp*.

The bottom-and-top box that was implemented resulted in several improvements:

- The box itself and the unit load (Figure 6.4) are much more stable, enabling safer and easier palletization at the producers. For some producers it also was easier to manually pack the product since there are "two" open boxes with the new solution.
- The new improved structural design improved stackability, enabling easer replenishment and a flexibility of warehouse display modes in stores (Figure 6.5). The previous solution had limited stackability and could only be transported vertically, with the boxes standing up. This also resulted in consumer difficulties when picking up and transporting the sofa from the store.
- Handles were added to further improve the design. *Ektorp* is heavy and bulky, making it difficult to handle. Adding handles enables two people to easily carry it, turning it into a customer-friendly package.
- Another important improvement was the use of straps instead of tape. The strap does not only look better but functions better than tape. For *Ektorp*, tape was a big quality problem. On a unit load or in a store self-service area,

Figure 6.4 The new, improved and implemented packaging system configuration containing three sofas.

(a)   (b)

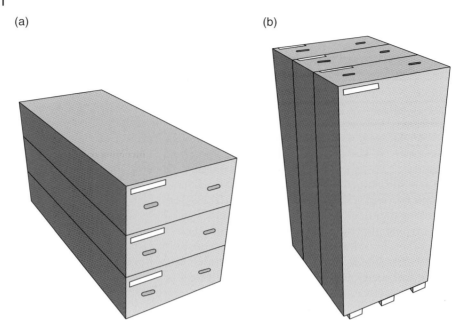

Figure 6.5 Illustrations of how the packages can be placed and stacked.

the tape from different boxes often got tangled, resulting in the boxes opening up. This in turn resulted in repacking activities, dusty products, and the like. With straps, no such quality problems have been reported.

Compared to the previous packaging solution, the trade-off for the bottom-and-top-box is that it requires more packaging material. This results in higher packaging costs. However, the supply chain profit is larger due to the advantages of the new packaging. From a planet perspective, one could argue that more packaging material is a trade-off. However, according to IKEA the planet is definitely the winner, since there are fewer damaged sofas and all the direct and indirect activities related to that have significantly deceased.

6.3.4   Concluding remarks: Knock-down boxing

This real-life case shows that packaging is closely related to product quality and that product protection is an essential direction in the packaging design compass for sustainable development. Even though the implemented solution requires more packaging material and costs, it improves the total impact on sustainability (Table 6.5). From a planet perspective, the benefits from reduction of product damages far exceed the use of more packaging material. From a people perspective, the new *Ektorp* packaging solution is superior. It enables easier and safer handling for both the producer and the retail store staff, and is easier for consumers to carry. From a cost perspective, the reduction of product damages along the supply chain (the indirect benefit of reduced workload and improved customer experience) far outweighs the increase in packaging cost.

Table 6.5 The impact (+ pros and – cons) on the three pillars of sustainability, of the redesigned *Ektorp* packaging solution.

	Planet	People	Profit
Producer	– More packaging material	+ Easier product packing + Easier palletization	– Increased packaging cost + More efficient packing operations
		+ Safer palletization	– Increased handling of incoming pallets
Transport	+ Less product damage		+ Less product damage
Distribution Centre	+ Less product damage		+ Less product damage
Distribution	+ Less product damage		+ Less product damage
Retail	+ Less product damage	+ Easier replenishment + Safer replenishment	+ Less product damage + Improved replenishment efficiency
			+ Improved display flexibility
Consumer	+ Less product damage	+ Easier carrying – More packaging material to recycle	+ Less product damage + Improved customer experience

Since all types of product-related damages are reported globally throughout its supply chain, IKEA has detailed information and insight into the magnitude of potential quality problems for each product. Across all product segments and with IKEA's enormous sales volumes, the handling and packaging-related costs of poor quality were over 200 million Euro for 2012. This indicates that there is room for improvement, both in the logistics and the packaging systems.

Case acknowledgements

Special thanks to the whole packaging development team at IKEA of Sweden. Very special recognition goes to Sigrid Svedberg, who acted as packaging development engineer in this redesign project.

References

Bertoluci G., Leroy Y. and Olsson A. (2013), Exploring the environmental impacts of olive packaging solutions for the European food market. *Journal of Cleaner Production*, 64(1), 234–243.

Christie G. (2001), *Controlled Release Grape Packaging*. Horticulture Australia Limited, Sydney.

Eriksson M. and Strid I. (2011), Livsmedelssvinn i butiksledet – en studie av butikssvinn i sex livsmedelsbutiker SLU Sveriges lantbruksuniversitet.

Food and Agriculture Organization (2002), Situation report and statistics for the world vitivinicultural sector in 2002.

Gunders D. (2012), Wasted: How America is Losing up to 40 Percent of its Food from Farm to Fork to Landfill, *NRDC Issue Paper.*

Hellström D. and Nilsson F. (2011), Logistics-driven packaging innovation: A case study at IKEA. *International Journal of Retail and Distribution Management,* 39(9), 638–657.

Jonsson A., and Kalling T. (2007), Challenges to knowledge sharing across national and intra-organizational boundaries: Case studies of IKEA and SCA Packaging. *Knowledge Management Research & Practice,* 5, 161–172.

Klevås, J. (2005), Organization of packaging resources at a product-developing company. *International Journal of Physical Distribution & Logistics Management,* 35(2), 116–131.

Klevås J., Johnsson M. and Jönson G. (2006), A packaging redesign project at IKEA. In: *Nordic Case Reader in Logistics and Supply Chain Management,* University Press of Southern Denmark.

Lindh H., Olsson A. and Williams H. (2016), Consumer perceptions of food packaging: Contributing to or counteracting environmentally sustainable development? *Packaging Technology and Science,* 29(1), 3–23.

Ras P.J. and Vermeulen W.J.V. (2008), Sustainable production and the performance of South African entrepreneurs in a global supply chain: The case of south African table grape producers. *Sustainable Development,* 17, 325–340.

Wallén A., Brandt N. and Wennersten R. (2004), Does the Swedish consumer's choice of food influence greenhouse gas emissions? *Environmental Science & Policy,* 7(6), 525–535.

Wikström F. and Williams H. (2010), Potential environmental gains from reducing food losses through development of new packaging – a life cycle model. *Packaging Technology and Science,* 23, 403–411.

Williams H. and Wikström F. (2011), Environmental impact of packaging and food losses in a life cycle perspective: A comparative analysis of five food items. *Journal of Cleaner Production,* 19(1), 43–48.

7   Material use

Material use is a direction that is present on all packaging, and is dependent on the product that is supposed to be packaged. In order to regard sustainable development holistically, you need to consider using a suitable material for the entire packaging system to fulfil its tasks. The material used, it features and the different technologies involved can contribute to more sustainable solutions.

In order to provide a basic understanding, this chapter starts with a general description of different packaging materials and their features. Examples of the consequences of using different materials and of combining them are presented and discussed. The focus is on food packaging and its effects on food shelf life. The reason for highlighting packaging material for food, is the sensitivity of food products as well as the high volume of food products produced and distributed.

This chapter then presents three cases. The first two cases illustrate the effects of using packaging materials that differ from the traditional ones. The first case compares the traditional retortable metal can, made out of one material in a circular shape, with the newer retortable laminated flexible carton-based squared packaging. This second case compares the traditional glass bottle used in the large wine industry in South Africa with the effects of introducing PET bottles for the same product. The third case illustrates and explores the effects of displaying the brand more clearly in retail by means of a suggested redesigned packaging system. The reduction of material is one of several factors in the redesign (two others being improved facing and less handling time), but they all play equally important roles in the performance of the entire system.

The cases are:

- Know and adapt your packaging material;
- Can or no can? The *Tetra Recart* retortable package;
- Wine in glass or plastic bottles;
- Facing the orange juice brand.

Managing Packaging Design for Sustainable Development: A Compass for Strategic Directions, First Edition. Daniel Hellström and Annika Olsson.
© 2017 John Wiley & Sons, Ltd. Published 2017 by John Wiley & Sons, Ltd.

7.1 Know and adapt your food packaging material

Case by Annika Olsson

Reports in the media along with government legislation call for minimizing the use of packaging material to achieve sustainable development. Consumers have responded to this around the world. In their perceptions and willingness to do the right thing for a better environment, their focus has turned to the packaging material when they select food products. In Sweden it is clear that consumers prefer paper and consider it to be the most environmentally friendly packaging material, and perceive plastic to be the least (Lindh et al., 2016). But is the packaging material really the major concern for sustainable development? This chapter aims to introduce the basic concepts for knowing your packaging material.

7.1.1 Food protection through packaging

As a packaging designer, you always need to consider the most important role of the package: to protect the product! This is especially important for food, since we put it into our bodies. The package's ultimate goal in the food industry is to protect its contents so that the food packed inside is delivered to the consumer in a safe and sound condition for consumption (Robertsson, 2013). The package needs to be tight so that it can safeguard the food it contains from undesired transfers of gases, microorganisms, light and other substances that speed up the deterioration processes. In this way, the product quality is maintained all the way to consumption.

Each food item has its own requirements for packaging material and differs in its sensitivity to the deterioration process from every other food item. Fresh food is the category that causes the highest percentage of waste, since deteriorating processes are still going in the package. For fruits and vegetables, the ripening process continues after harvest and for fresh meat the microbiological breakdown is still progressing. These processes depend largely on the transfers that occur between the surrounding environment and the internal environment of the package, as shown in Figure 7.1. Deteriorating processes are usually faster at higher temperatures and the product shelf life is thus dependent on both time and temperature. In addition, the natural presence of oxygen in many air-based food packaging environments hastens the chemical breakdown and microbial spoilage of the packed food.

These deterioration processes have an effect on resource utilization and the environmental burden in the production of different kinds of foods from raw material to ready consumable products. From a sustainable point of view, the preservation and protection offered by the package for food that generates a high environmental burden requires special attention. But while the foods inside packages may differ, the environmental load from the package itself generally constitutes only 3–5% of the total environmental impact in the food and packaging systems' life cycle.

The protection created through the package is essential in maintaining food quality and prolonging shelf life. The packaging system is constantly exposed to hazards from the external environment in storage, handling and distribution throughout the entire supply chain. By creating tight, resistant packages, package

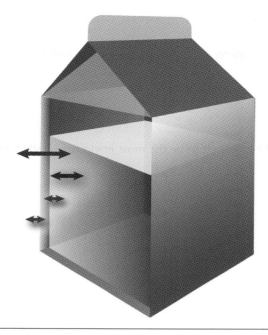

Figure 7.1 Undesired transfers between the surrounding environment and internal packaging environment.

integrity is created, but as soon as that integrity is broken the food is no longer preserved (Robertsson, 2013).

7.1.2 Different packaging materials

Package designers need to have in-depth knowledge of the deteriorating processes described above and of the different packaging material features that can control the rate of deterioration. Food protection is highly dependent on the protection ability offered by the package. This is mainly determined by the nature of the packaging materials selected when designing the package. There is a variety to choose from: glass, metal, paper and plastics are the most common sole materials, but combinations are also used to increase the protective properties of packaging. New materials are constantly being developed and recently, new bio-based polymers have been introduced. The interaction of these materials with different food products, though, has not been investigated and more knowledge on their protective properties is needed.

7.1.2.1 Glass

Glass is an inert material, originating from sand and does not interact with the environment or the food. Light is the one concern when using glass, since it shines through the material and can change the colour of light sensitive products. Another concern is the fragile nature of glass, which can lead to unnecessary product waste if broken under mechanical stress. The weight of glass compared to other packaging materials can potentially be seen as a drawback in handling and distribution.

7.1.2.2 Aluminium

Aluminium is a non-permeable material made from bauxite. It does not allow for transfers and can be considered the best barrier to all transfers including light. For acid and spicy products, there is a risk of interaction occurring between the product and the packaging material, and therefore aluminium might need to be protected by barriers of other materials. Both glass and aluminium require a considerable amount of energy in their production, although they are both fully recyclable.

7.1.2.3 Plastics (polymers)

Different kinds of plastics (polymers) usually have crude oil as their raw material, but the bio-based polymers of later development are made from crops and other natural sources. Polymers are all permeable to gases and light to different extents. The lower the permeability of the plastic material, the better the protection properties from molecules; the lower the transparency, the better the protection from light.

7.1.2.4 Paper

Paper is made of forestry products and is renewable in the sense that the forest regenerates after harvest. The best option is to use trees that grow faster than the rate of use. Cardboard is a good barrier to light, and to a certain extent to mechanical stress. But as a protective barrier, paper is extremely permeable to gases and small molecules and thus offers weak protection from deterioration. It is also sensitive to humidity, which is why it is often laminated with a protective barrier.

Even though packages made out of one single material are preferred, it is not always suitable. In many cases different materials need to be combined, since the product requires it for protective purposes. Combined packaging materials are made by laminating different materials together in order to fulfil the required protective properties. These packaging materials are made of different kinds of plastics laminated together, plastics combined with paper, plastics combined with aluminium or plastics combined with both paper and aluminium. One disadvantage for consumers is that packages made of these combined materials are harder to recycle than those made of pure materials because of the uncertainty of what recycling bin to throw them in. However, the increase in food protection is prioritized over ease of recyclability when selecting packaging material.

The packaging industry is constantly working to improve packaging materials and their protective properties. The aim is to reduce the environmental burden from packaging, and yet to maintain or improve its preservation and protective functions. One way to reduce the environmental burden of packaging is to reduce the amount of different materials used in a package. One example is when the fraction of aluminium used in flexible carton packages is reduced in favour of more paper. The protective properties of the different materials are then utilized to produce non-permeable, tightly sealed packages. But it is important for the packaging industry not to reduce the amount of packaging material too much. As explained earlier in this book it is better to have a package that is a little less environmentally friendly, as long as it better protects the product before consumption. This is because the waste of food due to deterioration results in an even greater environmental burden and resource waste than that of the mixed-material package.

7.1.3 Consumer preferences

Previous studies show that the dominating environmental concern that consumers have with packaging is related to its material features. They think the environmental burden of packaging is predominantly a matter of material. This reveals a great lack of knowledge, since in most cases the material represents a relatively small part of the environmental impact of packaging and has an even smaller impact on the total product-packaging system (Lindh et al., 2016). One reason for this lack of knowledge is that authorities and the media usually communicate that packages are to be regarded as waste. Another amplifying reason is that it is not until the package is empty and ready for recycling that consumers start to regard it as a package. Before that, it is one with the product it contains, and probably the reason why many consumers only view packages as waste. Hence, the major role of packaging in environmentally sound food production – which is to protect food from loss or waste – has a much greater impact than the packaging material itself.

7.1.4 Packaging technologies for increased shelf life

The selection of packaging material has large effects on the protective properties of the package and on the ability to seal them tightly. In addition, there are different food processing and filling technologies that further enhance the food shelf life. Food shelf life is thus a combination of the processing and filling technology, and the packaging quality. The suitability of the packaging material depends on the technology that is used.

The hot filling principle is common for cordials, marmalades and juices in glass bottles. This means that the product is filled at 100 °C, and when it enters the glass jar or bottle the container is sterilized by the hot product. This principle requires glass since most polymers are unable to tolerate the high temperature without the polymers migrating into the product.

Retort is another process and filling principle where the unprocessed food is first filled and sealed into the package, usually a metal can, and thereafter processed (sterilized) at a high temperature (121 °C) for a given amount of time in an autoclave system. This requires a material that can withstand high temperatures and the pressure imposed by the autoclaves. While the retort metal cans have

Figure 7.2 Principle for aseptic carton filling.

existed for more than 200 years, laminated carton materials have been developed recently to withstand these processes (see Section 7.2, "Can or no can? The *Tetra Recart* retortable package").

For other flexible carton systems, most often used for liquid food products, either aseptic principles or extended shelf life (ESL) principles are used in the processing and filling. This means for both principles that the product is heated and cooled down in a continuous process before filling. The packaging material is usually pre-treated with peroxide and UV light for sterilization before being formed into a package in which the processed product is filled and sealed. In aseptic technology, the packages are usually fed into filling machines continuously, formed like a tube and sealed through the product (Figure 7.2), and treated with higher concentrations of peroxide (for example) than for packages in the ESL principle. The processing and packaging material pre-treatment kills off the microorganism, resulting in the long shelf life of the product without any added preservatives.

The aseptic filling process means that the product is aseptic and can be stored in ambient temperatures for up to one year. For the ESL principle the treatment is milder and the shelf life is just extended compared to the ordinary shelf life of pasteurized products in chilled storage.

The main advantage with aseptic technology is that the products can be distributed and stored at ambient temperatures for a long time, which means they can be distributed over long distances and stored in different climates. The disadvantages, as with any heating processes, are that some vitamins and other quality parameters of the product decrease, compared to fresh products. The ambient conditions and long shelf life contribute to less product waste but may have some negative effects on taste and quality perceptions for the consumers.

Meat, fresh cut salads and fruits, and baked goods are usually packed in plastic trays with a transparent sealed lid or in plastic bags that are sealed with a headspace. In order to prolong the product shelf life, the traditional tray or bag packaging methods with air in the headspace are often replaced by Modified Atmosphere Packaging (MAP). With the MAP technique, the air with the natural oxygen content inside the package is replaced with other gases such as carbon dioxide (CO_2) or nitrogen (N_2) to achieve different and more beneficial relations between the three gases (O_2, CO_2 and N_2). The gases are used in different products to create a gas environment inside the package that has a reduced level of oxygen and for some products a completely oxygen-free environment for pro-longed shelf life. With the MAP technology, the shelf life of a package of ground meat is often increased from one day to eight days, although this varies depending on the quality of the meat; but MAP can also be used to further enhance the appearance of the product. For that purpose, the level of oxygen is increased to about 80% in a meat package, which results in the meat retaining is red coloured throughout its shelf life. The disadvantage of MAP is that the meat looks better but is less tender. MAP can be introduced in different gas combinations, depending on whether the purpose is only to prolong shelf life or to enhance appearance.

By using a complete MAP, that is, by completely replacing the oxygen with other gases, an anaerobic environment is established inside the package (Lewander et al., 2009). The advantage is that spoilage microorganisms that need an aerobic environment in which to grow will be reduced and the spoilage process will slow down. The disadvantage is that anaerobic pathogenic microorganism can start growing. This can happen for some food products but not for all. These pathogens may not cause overt evidence of spoilage but can cause harm to the consumer when consumed. The gas as such does not improve the quality of the meat but rather prolongs shelf life since it stops the deterioration processes.

The introduction of gas into food packages has been debated, mainly concerning meat. This is most likely because it has both positive and negative impacts on the perceived meat quality. For other products packed in a modified atmosphere, such as baked goods and fruit juices, the consumers seem to be unaware of the fact, even though regulations require that the information "packed with protective gas" appears on the package.

The clearest gain with MAP from a sustainability point of view is its ability to prolong the shelf life of sensitive food products. This potentially results in less waste of fresh produce, particularly in the homes of the consumers. But the technology is also frequently used for other products, such as fruit juices that are packed with an N_2 headspace in order to reduce oxygen and prevent the juice from "browning" through oxidation. For such products the best-before date can be moved forward, and the shelf life increased, but the effects of less waste in these cases are not as evident. From a sustainability point of view, longer shelf life can increase the amount of storage that is required, increase the risk that older products will end up at the back of the shelves, and that consumers will intentionally pick the products with the longest shelf life and not the older ones so that they end up going to waste. All such effects need to be balanced into the waste discussion of whether MAP actually results overall in less waste or just moves the problem to another actor in the value chain (Table 7.1).

Table 7.1 The impacts (+ pros and – cons) on the three pillars of sustainability, of selecting protective food packaging material.

	Planet	People	Profit
Producer	+ Better protected food		+ Less food waste
	+ Less food waste		– Increased packaging cost
	– More packaging material		
Transport	+ More stable packages	+ More stable packages, enables handling	+ Fewer transports
Distribution Centre	+ More stable packages	+ More stable packages, enables handling	
Retail	+ Better package appearance	+ More stable packages, enables handling	+ Less food waste
	+ Less product waste		+ More sales
Consumer	+ Longer shelf life at home	– May make less informed choices leading to more food waste	
		– Concerns about protective gases	
	– Less packaging in recycling due to lack of knowledge	+ Awareness might change behaviour	

7.2   Can or no can? The *Tetra Recart* retortable package

Case by Annika Olsson

The canning of food started during the Napoleonic Wars when food was sterilized in glass jars. Tin plate cans were then developed and introduced when the British Empire expanded all over the world (Beckeman and Olsson, 2005). In-container sterilization at high temperature and under pressure is called retorting. The container is first filled with the unprocessed food, sealed and then sterilized in an autoclave system. Sterilizing food in a completely sealed and tight metal container provides long shelf life for a safe food product that can be stored and transported at room temperature. Naylor (2000) emphasizes the impact the can has had in the statement: "Indeed, the canning of food was a decisive moment in the growth of globalization." This is confirmed by a presence of global products in cans and the huge size of the worldwide market of cans.

7.2.1   The package configuration and the redesign

It has taken about 200 years of packaging development to come up with a new package solution for conserving food that provides long shelf life in a way similar to that of retortable cans or glass jars. The *Tetra Recart* package, developed by Tetra Pak, is the first carton-based package where food is sterilized in the package after filling.

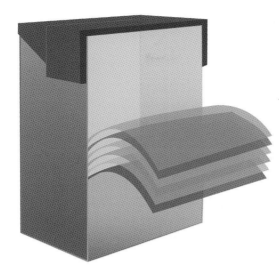

The principle laminate setup
• Inside polypropylene layer
• Aluminum foil
• Polypropylene
• Paperboard
• Outside polypropylene layer

Figure 7.3 The material set-up in a *Tetra Recart* package.

This innovation was awarded the prestigious Marcus Wallenberg Prize in 2005. The carton package is specially designed for food products traditionally packed in retort cans or glass jars. The first test of the *Tetra Recart* was carried out in close cooperation with the Pet Food Division of *Nestlé* in 2001. In 2002 it was tested and launched with the *Bonduelle* brand in France for conserved vegetables (Higgins, 2004), even though the initial development of the new material started as early as 1983.

From a technical perspective, the development focused on finding a packaging material that could withstand the pressure and heat up to 130 °C as well as 100% relative humidity from the process for up to two hours without interfering with the packed food and without affecting the packaging stability. The package consists of a multilayer laminate with paperboard as the main component (Figure 7.3).

The principle laminate set-up is:

• inside polypropylene layer
• aluminium foil
• polypropylene
• paperboard
• outside polypropylene layer
• printing and coat of lacquer,

Apart from the packaging material development, consumer benefits were also considered, such as a hand-friendly shape and no sharp edges after opening. To enhance user-friendliness and make it easier to open for consumers, the package has a laser perforation on the top flap (Olsson and Larsson, 2009) (Figure 7.4). Different products are packed and sold in the new carton, but similar products are still available in a tin or aluminium can at slightly different prices for the same product in the same retail outlet in Sweden. For example, the prices of the same chickpeas product in the two packages in Figure 7.4, were about 15% higher per kilo for the can than for the carton package. The price difference is most likely

Figure 7.4 Chickpeas packed in a metal retort can and in a flexible retortable carton package.

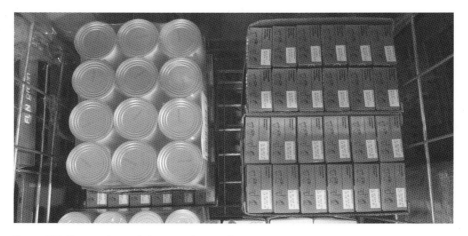

Figure 7.5 Cans and cartons in secondary packages.

related to both the package type and the product as such, since the two products are different brands.

The two types of packages have different environmental profiles, both in the material as such and in their shape and stackability features. Obviously, a circular package will take up more space in the secondary package, even though both packages are usually transported as 12 primary packages in a secondary package. Figure 7.5 shows one tray of twelve cans placed on top of two trays of carton packages (24 packages) to illustrate the space utilization of the different package

Table 7.2 LCA figures for different primary packages and their space utilization in the packaging system.

Components	*Tetra Recart*	Tinplate can	Aluminium can	Glass jar
Weight primary package (g)	18.2	54.6	24.0	227.2
Tray + wrap foil weight (g)	66.7	55.1	55.1	58.9
Packages per tray	12	12	12	12
Trays per layer	17	13	13	12
Layer per pallet	9	8	8	8
Primary packages per pallet	1836	1560	1560	1152

types. It also shows the different secondary packages, where the cartons most commonly are packed in cardboard boxes, while the metal cans usually have a cardboard tray with a wrap-around plastic cover.

In a Life Cycle Assessment (LCA) analysis carried out in Germany in 2004, the flexible *Tetra Recart* carton, a tin can, an aluminium can, and a glass jar with the same product volumes were compared. The LCA analysis reported the measurements (Kruger et al., 2008) shown in Table 7.2.

The comparison shows that a single can weighed more than a *Tetra Recart* carton, while the weight of the tray was slightly higher for the carton package. Meanwhile, the tray is more efficient for cartons because of the squared shape of the primary package. This means that you are able to fit more trays on each layer of the pallet.

7.2.2 The comparison – sustainability implications

The main and foremost differences with the carton package are in the shape and weight of the material compared to metal or glass packaging. The material changeover is clearly radical in many aspects. The weight is approximately one-third of the medium weight of a similar sized tin can; for the aluminium can, the difference is not as large but still significant.

The difference in the environmental impact is obvious with the relatively lower weight of the carton package compared to the metal cans, not just because of the material but also because of the different weights in transport load. In addition, the main portion of the weight in the carton-based package is made up of paper, a renewable source of material. On the other hand, the recycling rates of tin and aluminium are higher than for cartons and the recovery of material is also higher for the metals (Kruger et al., 2008) (Table 7.3).

Even though the material as such results in clear differences in the environmental performance of the two types of packaging, the most dramatic change is the effect of the new shape on the fill rate. The squared shape, which is made possible because of the flexibility of the laminated carton material, makes the solution more space efficient in transport and storage, as well as in the display on retail shelves. The *Tetra Recart* provides better logistical performance in the

Table 7.3 Collection rates (in recycling systems) of different packaging materials for retort packaging.

Components	Tetra Recart	Tinplate can	Aluminium can	Glass jar
Collection rate % (Germany)	70	85	85	85
Recovery at sorting %	89.5	95.9	95.9	97.5
Overall recycling rate %	62.5	81.5	81.5	82.9

Figure 7.6 The principle difference in transports (inbound and outbound from production) of retortable cans compared to *Tetra Recart*.

outbound, from production to consumption, because a rectangular package is more space efficient than a circular one. The major logistical advantage is in transport and logistical activities prior to filling and processing, that is, in the handling of empty packages inbound to the producing factories (Olsson and Larsson, 2009). This is because the carton-based package is transported in a folded flat condition, while cans are transported as empty containers resulting in the transport of more "air". This is a clear material advantage where the laminated carton is flexible and able to be supplied as flat blanks to the filling machines (Figure 7.6).

From an environmental point of view, the difference in the overall number of transports is evident, with a decrease of about 55% (Olsson and Larsson, 2009) for the *Tetra Recart*. Since the retortable can market is large (more than 160 billion cans annually) and the type of package allows for long distance distribution (the product can be stored for a long time and transported under ambient conditions), a substantial impact on the environment (planet) as well as on the economy (profit) would be achieved if cans where replaced with cartons. Due to the structure of the carton packaging material, there may be a higher risk of waste or damaged cartons compared to metal cans, since the carton package is more sensitive to mechanical stress. But with well-constructed secondary packages, this risk can be minimized. The square shape of the primary package already allows for tighter stacking in the secondary package, resulting in more protection of the sides of the primary package. In developing countries, where packages are stored on the ground and sold in open markets, the reduced protection from mechanical stress as well as from animals such as rats needs to be taken into even greater consideration. In certain environments, the higher mechanical robustness of the metal can rule out the benefits of the carton-based package.

The economic impact of replacing tin cans with carton packages is most likely less than the environmental impact. This is because replacement of the existing filling equipment and can production lines is costly to food manufacturers. The investment costs may be a reason to install a modest base of retort carton packages in relation to the existing base for metal cans. Another reason for the limited economic impact can be the force of user habits and traditions for the old can. However, it is worth noting that the end consumer price is lower for the carton package than for the can with the same product in the supermarket, even though they are not directly comparable since they contain two different brands. But it is evident that the retailer (who has its own brand packed in carton packages) affects consumer choices with a lower price.

Apart from the difference in environmental impact from weight and transport, the shape of the package also offers economic advantages with fewer transports and more efficient stacking in the secondary packaging. Handling times in the different parts of the supply chains also decrease with the square-shaped solution and less weight, resulting in cost reductions. The shape allows for better exposure and better communicative opportunities for brand owners (Olsson and Larsson, 2009), an aspect that can increase sales, market shares and thus profit for the brand owner. The disadvantage of a weaker package, from a mechanical stress point of view, is that its appearance on the shelf can be negatively impacted, which results in unsold packages and an increase in waste.

From a social perspective, the package is constructed to be more convenient for the end user with the laser perforation on the upper flap. It allows the consumer to open the box without additional tools, although the complexity of the material and the tight seal may still cause problems for people with weak hands. Another people aspect is the reduction in personal injuries from the sharp edges of the metal material in cans (Higgins, 2004). On the other hand, the can has a secondary value in developing countries, where it can be used for cooking when emptied and for storing food.

From a recycling perspective, the carton package is easier to fold before recycling, but it has the disadvantage of being made of a multi-material set-up. Although there is a recycling system for cartons, the multi-material structure may confuse consumers as to where to put the empty package, with a risk of more cartons packages ending up in the wrong bin or not being sorted or returned at all. A metal can is easily recycled, since it has a one-material construction.

An evaluation of the retortable can compared to the metal can is made in Table 7.4.

Table 7.4 The impact (+ pros and – cons) on the three pillars of sustainability, of the retortable carton package.

	Planet	People	Profit
Producer	+ Efficient to store packaging material	+ Less weight and easier handling in filling procedure	+ Less storage space for empty packaging
			– Investment in new packaging solutions and equipment
Transport	+ Efficient stacking, high fill rate	+ Better stackability less handling, fewer returns	+ More efficient transport
	+ Less weight		– Less protected, weaker package
	– More waste from destroyed packages?		
Distribution Centre	+ Efficient stacking and better space utilization	+ Better stack-ability less handling, fewer returns	+ More efficient handling
Distribution	+ Efficient stacking, high fill rate	+ Better stack-ability less handling, fewer returns	+ More efficient transport
	+ Less weight		– Less protected, weaker package
	– More waste from destroyed packages?		
Retail	+ Less secondary packaging to recycle	+ Less secondary packaging to handle	+ Better facing on shelves
	+ Shelf-ready secondary package with brand name	+ Lower weight of packaging system (primary and secondary together)	– Poorer package appearance if the package is damaged
Consumer	+ Easier to flatten prior to recycling	+ Fewer injuries	
		+ Easier to open	
	– Fewer packages recycled	+ Easier to store at home	

Case acknowledgements

Special thanks to the product manager of *Tetra Recart*, Helena Brändström, for sharing press releases and other media material about the *Tetra Recart* packaging system.

7.3 Wine in glass or plastic bottles

Case by Fredrik Nilsson

Wine is a fast-moving consumer good that is sourced and moved around the globe. The glass bottle is the traditional packaging for wine. However, due to its fragility, costs and environmental impact, different wineries are looking for new alternatives of primary packaging (e.g. carton based or plastic). This case elaborates the question of using glass or plastic bottles for wine. The case was conducted in South Africa at a winery that exports most of what it produces to Europe and the US. The aim was to explore how the winery could most effectively reduce the CO_2 levels that arise due to packaging and packaging logistics activities.

In South Africa, the wine industry plays an important role in the agricultural sector. South Africa's annual wine harvest produces a billion litres of wine. There has been an increase in the last few years going from under 300 wineries in 1997 to almost 600 wineries in 2010 (Vink et al., 2012). Out of the top ten wine producing countries in the world, South Africa is eighth (The Wine Institute, 2014). To retain its place in the top ten and to ensure its competitive advantage, one area that the South African wine industry is looking for improvements is in the area of packaging (Chance, 2010). For example, the Packaging Council of New Zealand advocates maximizing transport efficiencies by lowering the transportation of "air" in packages. The decrease in air increases the fill rate of the packaging system. This means fewer transports and lower carbon emissions in logistics.

7.3.1 The packaging system

The studied winery uses glass as its primary packaging, with cardboard cases and mesh as secondary packaging. Each 750 ml bottle used in this case measures 80 mm in diameter, with a height of 298 mm and weighs on average 860 g when empty. A full bottle of wine weighs more than 1.5 kg. Hence, a box of wine with 12 bottles weighs 19 kg. The tertiary packaging level consists of the finished boxes of wine that are packed onto a pallet and then wrapped with shrink wrap. The boxes are stacked 4 layers high with 14 boxes on each layer. The shrink wrap provides extra stability during transportation, handling and storage. For transport to Europe, Euro pallets made of wood are used, which makes them quite heavy. The advantage is that the pallets are very durable and strong, which is necessary when handling heavy bottles of wine. The pallets can also be handled easily by using a forklift.

This case focuses on the primary package: the wine bottle. The other levels of the packaging system have remained the same. More specifically, one of the rosé wines was studied from the bottling process until reaching the consumer in Europe, in order to determine the consequences a change from glass to PET (polyethylene terephthalate) would have on the supply chain, especially concerning the environmental impact.

7.3.2 Supply chain description

The winery receives its grapes from selected farms around the Stellenbosch area. The grapes are harvested on these farms and transported to the winery for processing and fermentation. The grapes undergo various stages of preparation before being bottled and packaged. The wine is bottled at the winery and no third party is involved in this step. The bottles are steamed and washed by a machine to ensure that they are clean and sterilized. After that they are blasted with warm air to dry, which is performed automatically. The bottling is semi-automatic and starts when the cleaned wine bottles are placed on a conveyer belt by one of the employees in the store. All the wine bottles are then packed in cardboard boxes (secondary packaging) in groups of twelve. Once a box is full, it is folded closed and sealed with heavy duty tape. These boxes are placed onto pallets (tertiary packaging), stacked on top of each other and stored until they are to be transported. When the wine arrives at the distribution centre it is placed in storage until orders are received from various customers. Once the orders are received, the pallets are broken down and new mixed pallets are made up for each specific order and placed on a truck. The new pallets are then transported to the harbour and from there shipped to Europe or the US.

7.3.3 From glass to PET

It is a matter of fact that the glass bottle is and has been the dominating packaging for wine. Due to both tradition and consumer preferences, the glass bottle is associated with high quality while other packaging material is still considered to be the budget choice. However, due to consumer changes, the box-wines are growing globally and the wineries are also trying to find both more cost-effective and environmentally friendly solutions. While glass is a 100% reusable resource that can be recycled countless times without any degradation of the raw materials used, plastic alternatives provide other features. For example, PET has the necessary properties required to bottle wine. It also boasts a reasonable barrier to gas and moisture, alcohol and solvents. It is semi-rigid, depending on the thickness. Coloured PET provides a reasonable barrier against sunlight, a prerequisite in the wine industry. Its versatility means that it is ideally suited to packaging purposes because it offers the required protection against environmental factors, as well as against physical damage. This translates into an extension of the product's life cycle, and less secondary protective packaging than that required for glass. Furthermore, as concluded by Pattara et al. (2012), packaging, primarily the glass bottle, counts for the greatest contribution in terms of emissions (more than 70%), followed by product distribution and agricultural operations.

The cost of producing one glass bottle, according to the production manager at the winery, is SEK 8.54 (approximately €1) based on the different costs per component (0.62 for the wire, 1.40 for the cork, 0.60 for the cap and sleeve, 0.55 for the neck and label, €0.26 for the back label, 0.26 for the front label, and 4.85 for the glass bottle). Based on figures from a PET supplier, the cost for one bottle of PET is SEK 2.32 (approximately €0.25) (0.10 for the cap, 0.26 for the back label, 0.26 for the front label, and 1.70 for the PET bottle). Hence, the packaging cost per bottle of wine is 3.9 times more expensive for a glass bottle than a PET bottle.

Arena et al. (2003) conducted a life cycle assessment of the recycling system for synthetic polymers (plastic) in Italy. The study found that to produce 1 kilogram (kg) of recycled PET (R-PET), a gross energy ranging between 42 and 55 megajoules (MJ) was required. To produce the same quantity of virgin PET, more than 77 MJ were needed. From a cost perspective, the cost benefit of recycling glass in the UK is about €0.21, while PET is about €8 per ton of material (Craighill and Powell, 1996).

By using the emissions factors from the Wuppertal Institute of $0.716\,g\,CO_2$ per g of virgin glass, the 860 g glass bottle used for the South African wine emits 618 g of CO_2 emissions during the production stage. The Association of Plastic

Manufacturers has gathered data on the quantity of energy that needs to be generated to produce one PET bottle. The data show that the average amount of carbon emissions that arise from manufacturing one virgin, 54 g PET bottle is 222 g of CO_2. This data includes and refers to all emissions that arise throughout different phases of production (WRAP, 2007).

7.3.4   Other packaging alternatives

While this case has focused on glass vs. plastic, there are a number of alternatives for products like wine that we have seen on the market for some time: the carton-based bottles and the bag-in-box solutions. Since empty carton bottles are supplied in rolls, one truck of empty cartons equates to 26 trucks of empty glasses. This means that transportation before filling alone results in a huge reduction in carbon emissions. Added savings for the producer are that no corks, foils or additional labels (needed for bottles) are required. Carton packages are cheaper than glass and can be distributed more efficiently once full. They are easy to stack and more of them can be transported in one load than glass. The bag-in-box solution (1.5–5 litres), and the new version with only the plastic bag, is another type of packaging that is more efficient and environmentally friendly when it comes to handling and transportation.

7.3.5   Implications from using PET wine bottles

There are several implications from a people, profit and planet point of view in this case (Table 7.5). When it comes to transportation and distribution, the difference of 806 g in the weight of the two bottles impacts the costs as well as the environment. Hence, distributing 10,000 bottles would result in an 8.06 ton reduction of weight to be handled and transported. While the volume efficiency would be the same, the weight would reduce the fuel consumption, having both cost and environmental impacts. Furthermore, the total amount of CO_2 would decrease from 6.18 tons to 2.22 tons for the same number of bottles. Thus, implementing PET bottles for wine will affect both the planet and the profit.

The key findings in this case were that the implementation of PET bottles could save the winery a significant amount in transportation costs, while simultaneously reducing its carbon footprint. If the change to PET bottles was implemented by the winery, there is a potential for more positive effects on its entire supply chain. The advantages include a decrease in the weight of transportation and in a bottle's production cost of €0.67, while simultaneously reducing CO_2 by nearly a third from 620 g per glass bottle to 220 g per PET bottle. Even if glass wine bottles were manufactured using 100% recycled glass, the carbon footprint for manufacturing would only drop to about 295 g of CO_2 per bottle, which is still about 1.4 times that of virgin plastic bottle manufacturing (Neil and Michael, 2008).

Nonetheless, due to long-standing wine traditions, switching to alternate, more efficient packaging systems will be difficult. The average consumer is resistant to change, and coupled with the centuries-old traditions surrounding wine, there are bound to be a lot of problems when you change the *status quo* of its packaging. However, as consumers are becoming more environmentally concerned, the acceptance of other packaging alternatives for wine is growing. The growth of

Table 7.5 The impact (+ pros and – cons) on the three pillars of sustainability, of PET in the wine supply chains.

	Planet	People	Profit
Producer	+ Less packaging material + Less warehouse space needed	+ Easier handling in receiving, production and handling	+ Lower packaging material costs – Investment in equipment
Transport	+ Increased fill rate		+ Lower transport costs
Distribution Centre	+ Fewer broken packages	+ Easier picking + Easier handling	
Distribution	+ Increased fill rate		+ Lower transport costs
Retail	+ Fewer broken packages	+ Easier handling in storage	+ More efficient refill processes
Consumer	– Doubts about the material – Shorter shelf-life	+ Easier to handle	– Fewer sales

bag-in-box wines on the US market was 160% from 2003 to 2005 and even premium wines were accepted and sold in that way in 2005 (Santini et al., 2007).

Another reflection that can be made from this case concerns the use of natural corks, plastic corks or screw caps made of metal. Most people are under the impression that the cork trees are cut down to process the cork, but this is not the case. Instead, the outer bark from the trees is harvested by hand every nine years. This allows for the tree to consume 10 tons more carbon dioxide. Recent studies have shown that CO_2 emissions resulting from the life cycle of a screw cap are 24 times higher than those from natural cork, while the plastic stopper is responsible for 10 times more CO_2 than natural cork. Cork is thus a more viable option if you want to reduce your carbon footprint (Fisher, 2010).

Case acknowledgements

Special thanks to Alexander Muller for contributions to the case.

7.4 Facing the orange juice brand

Case by Annika Olsson

The global liquid drink consumption is estimated to be about 5,000 billion litres a year, of which about only 1,000 billion are packed. Slightly more than 100 billion are packed in carton-based packages and about 30 billion of those contain juice, nectars and soft drinks.[1] In this segment, plastic bottles compete with cartons, accounting for about 60 billion litres of juice, nectars and soft drinks. Evidently, the drive of the packaging industry is to increase the market share and sales of the different packaging systems.

This case[2] evaluates a redesign of secondary and tertiary packaging in a packaging system for orange juice sold in 1 litre aseptic laminated carton packages. The new design was also intended to cover other primary package sizes (200 ml, 250 ml and 500 ml), but the evaluation was carried out on the 1 litre cartons.

7.4.1   The package configuration and the redesign

Traditionally, aseptic cartons are sold in cardboard trays with 12 packages per tray. In this case, the 12 packages were 1 litre cartons of orange juice of a particular brand. The trays are currently placed on a half-sized standard wooden Euro pallet and often sold directly from the pallet, which is placed on the floor in the retail location.

A new pallet made of plastic for tertiary packaging was designed, in order to make the packaging system for the 1 litre juice cartons more flexible. It was a quarter the size of a full standard wooden Euro pallet. The plastic pallets could be connected to each other to build a half- or full-size pallet.

Instead of the traditional glued cardboard tray, a flat sheet of cardboard was used to cover each layer of the packages with a "look-in" solution of transparent material that was folded around the rims and corners of each layer. The new solution can keep the packages in place on the layers and protect the corners and rims of the packages on the pallet. The traditional tray solution was stacked in a cross-sectional overlapping pattern, with the front of some trays and the long side of others facing the consumer. The new layer solution allowed for the front of all primary packages on the pallet to face the consumer (Figure 7.7).

7.4.2   The comparison of the packaging systems

A comparison was made between the traditional Euro pallet and the newly-developed solution with the one-quarter size plastic pallet and the new cardboard layers.

The first and most straightforward comparison was of how much material was used on the secondary packaging level. The new redesigned solution resulted in a total reduction of 65% of the cardboard used compared to the traditional 12 × 1 litre trays. The handling time spent by personnel at the wholesale and retail locations was also measured as a secondary effect of the 65% material reduction. In order to compare the numbers, the calculation was measured in handling time per litre of orange juice. The results are shown in Table 7.6.

---

1 Tetra Pak statistics from 2011
2 This case has been published in another version in *Packaging Technology and Science* (2002): "Packaging throughout the Value Chain in the Customer Perspective Marketing Mix" (Olsson and Györei, 2002)

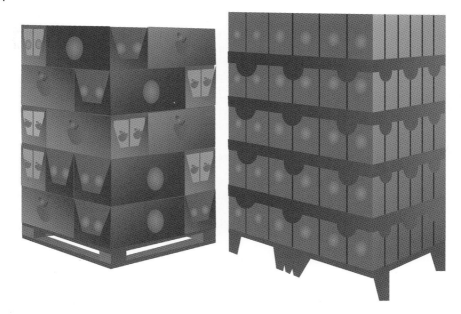

Figure 7.7 Sketches of the two designs: the traditional tray solution on the left and the redesigned layer solution on the right.

Table 7.6 Handling time per litre (sec/litre) (Olsson and Györei, 2002).

	Redesigned one-quarter plastic pallet	Traditional half-size Euro wooden pallet
Wholesaler	4.2	34.2
Retailer	4.8	35.0
TOTAL	**9.0**	**69.2**

The difference in handling time is substantial: 69.2 seconds were spent per litre of product in the traditional solution, while only 9.0 seconds were spent per litre in the new solution.

One hypothesis was that the system's new layout would attract consumers and result in increased sales. To estimate differences in sales volumes, the new and the old solutions with exactly the same products were placed in retail locations in four countries: Sweden, the UK, Russia and Estonia. The purpose of the tests was to determine how attractive the two packaging solutions were to consumers. The same brand name was displayed in both of the retail stores at two different locations:

1) "the market place", where the two different pallets were placed side by side directly on the floor in an area where other similar products were sold; and
2) "on the shelf", where the primary packages in trays from the traditional solution were put on a shelf, and the redesigned pallets were put on the floor next to that shelf.

Table 7.7 Evaluation on a 7-point scale of packaging features (Olsson and Györei, 2002).

Retail location/value	Marketplace		Retail Shelf	
	Redesigned layer unit ¼ plastic pallet	½ standard wooden Euro pallet	Redesigned layer unit ¼ plastic pallet	½ standard wooden Euro pallet
Accessibility	6.4	3.8	6.1	4.9
Visibility	6.7	3.8	6.0	5.5
Attractiveness	6.2	3.5	5.5	5.3
Quality	5.2	3.3	4.4	4.9
Preferable alternative	98%	2%	65%	35%

The value rating from consumers was based on quantitative studies in the four countries complemented by 100 consumer interviews on the Swedish market. In the quantitative studies, calculations of the number of packages consumers picked from the two units were made. The results showed a potential sales increase of 16% in Estonia, 11% in the UK, 15% in Russia, and a significant increase of 65% in Sweden in the packages picked from the redesigned pallets. In the 100 consumer interviews carried out in Sweden, consumers were asked to rate the two sales solutions on their perceptions of four different parameters: accessibility, visibility, attractiveness and quality. They graded the parameters on a 7-point scale, where 1 was low and 7 was high (Table 7.7).

The results from the consumer study clearly show that when placing the two solutions side by side on the floor in the retail location, the consumer preferences for the new solution were high. The comparison did not show the same significant difference when the traditional trays were placed on the shelf, although there was still a preference for the floor-based layer solution.

The new unit also proved to be easy to locate and handle in the stores due to its shape, size and format, although this was not explicitly measured.

7.4.3 Sustainability implications

From a sustainability perspective, the 65% reduction of cardboard material in the secondary packaging compared to the layer design obviously results in less environmental strain because less material is being consumed. From this perspective alone the redesign is sustainably favourable, thus good for the planet. Less material is also less expensive, meaning lower costs for the producer and a more economically sustainable solution. A secondary effect is that less material means less handling time for the people working with replenishment in retail. The new solution reduced handling time by a total of 85% in retail and at the distribution centre. In this way the new solution has positive effects from the perspectives of people, profit and planet.

From the manufacturer's point of view, however, the new system would require an investment in new machinery for efficient packaging, leaving the producer

Table 7.8 The impact (+ pros and – cons) on the three pillars of sustainability, of the redesigned plastic pallet solution.

	Planet	People	Profit
Producer	+ Less packaging material	+ Less packaging material to handle	+ Innovative customer solutions – Investment in new packaging solutions and equipment
Transport	+ Less packaging material	+ Less packaging material to handle + Easier to identify product	– Less protected primary packages possibly leading to more damaged packages
Distribution Centre	+ Less packaging material + Easier to see product, fewer mistakes	+ Less packaging material to handle	+ More efficient handling
Retail	+ Less packaging material	+ Less packaging material to handle + Fewer injuries in handling secondary packaging	+ Less packaging material to handle + Increase in sales + Less handling time
Consumer		+ Easier to pick product from pallet + Easier to identify brand	

with new investment costs. The increased investment cost was the main reason for NOT pursuing this development and which is why the new packaging solution never reached the market. The case shows the difficulty in calculating estimated sales increases and comparing this to the investment costs that are usually easier to estimate.

From a supply chain perspective, all actors seemed to benefit from the material reduction economically, both in terms of reduced material costs and reduced personnel costs in packaging handling. Less material in all steps of the supply chain is also beneficial from an environmental point of view, as long as it does not result in increased damage to the primary packages (Table 7.8).

7.4.4 Concluding remarks: Facing the orange juice brand

Since the new packaging system was never realized, one can argue from a sustainability point of view that the economic sustainability parameter (profit) needs to be evaluated throughout the entire value chain and not only from the perspective of one actor, the producer in this case, being responsible for the investment costs. To attain sustainable solutions, one can also argue that costs as well as profits require new business models where investments risks are shared among the actors in the supply chain.

Given the positive effects from both the planet and people perspectives in terms of reduced material and reduced handling times, measurements to balance these parameters were not made in the evaluation of the new packaging system. For example, measuring the amount of destroyed or damaged primary packages due to less protection and handling aspects could potentially have reduced the significant positive numbers for cardboard reductions. And if the new solution resulted in more damaged packages, package appearance would most likely have reduced the consumer acceptance of purchasing packages that were not intact. Such effects were not taken into consideration in the evaluation of the new packaging system. This shows the importance of evaluating new systems holistically on all three parameters of profit, people planet, and that the parameters are interlinked.

Case acknowledgements

Special thanks to Michel Györei for sharing input from this development case in the Distribution Department at Tetra Pak.

References

Arena U., Mastellone M. and Perugini F. (2003), The environmental performance of alternative solid waste management options: a life cycle assessment. *Chemical Engineering Journal*, 96, 207–222.

Beckeman M. and Olsson A. (2005), Driving forces for food packaging development in Sweden. *World Food Science*, pp. 1–15: http://www.worldfoodscience.org/cms/

Chance K. (2010), *Plastic bottles for wine are not inferior*. BusinessDay: http://www.businessday.co.za/articles/Content.aspx?id=114620 Accessed 30 November 2015

Craighill A. and Powell J. (1996), Lifecycle assessment and economic evaluation of recycling: A case study. *Conservation and Recycling*, 17, 75–96.

Fisher D. (2010), Cork vs. screw cap debate goes environmental: http://www.thinkgreenliveclean.com/2010/08/cork-vs-screw-cap-debate-goes-environmental/ Accessed 30 *November* 2015.

Higgins K.T. (2004), Stick-to-it-iveness defined. *Food Engineering Magazine*. 16 November.

Kruger M., Detzel A., Herová M. and Mönckert J. (2008), *LCA for Tetra Recart and alternative packs, status report 2005/2006*, IFEU GmbH, Heidelberg.

Lewander M., Lundin P., Svensson T., Svanberg S. and Olsson A. (2009), Non-intrusive measurements of headspace gas composition in liquid food packages made of translucent materials. *Packaging Technology and Science*, 24(5), 271–280.

Lindh H., Olsson A. and Williams H. (2016), Consumer perceptions of food packaging: contributing to or counteracting environmentally sustainable development? *Packaging Technology and Science*, 29(1), 3–23.

Naylor S. (2000), Spacing the can: Empire, modernity, and the globalization of food. *Environment and Planning*, 32, 1625–1639.

Neil J. and Michael R. (2008), *Comparing the carbon footprint of plastic and glass wine bottles*: http://wineenabler.com/comparing-the-carbon-footprint-of-plastic-and-glass-wine-bottles/ Accessed 12 April 2012.

Olsson A. and Györei M. (2002), Packaging throughout the value chain in the customer perspective marketing mix. *Packaging Technology and Science*, 15, 231–239.

Olsson A. and Larsson A.C. (2009), Value creation in PSS design through product and packaging innovation processes, Chapter 5, in: *Introduction to Product/Service-System Design*, Sakao and Lindahl (eds), Springer, pp. 93–108.

Pattara C., Raggi A. and Cic A. (2012), Life cycle assessment and carbon footprint in the wine supply chain. *Environmental Management*, 49, 1247–1258.

Robertson G.L. (2013), *Food Packaging Principles and Practice*, 3rd Edition. CRC Press Taylor and Francis, Boca Raton, FL.

Santini C., Cavicchi A. and Rocchi B. (2007), Italian wineries and strategic options: The role of premium bag in box. *International Journal of Wine Business Research*, 19(3), 216–230.

Vink N., Deloire A., Bonnardot V. and Ewert J. (2012), Climate change and the future of South Africa's wine industry. *International Journal of Climate Change Strategies and Management*, 4(4), 420–441.

The Wine Institute (2014), *World Wine Production by Country 2014*: http://www.wineinstitute.org/files/World_Wine_Production_by_Country_2014_cTradeDataAndAnalysis.pdf

WRAP (2007), *The Food we Waste*. Report. http://www.wrap.org.uk/

8 Fill rate

The direction fill rate is about space utilization. It refers to volume and weight efficiency. To achieve the best fill rate means to find an arrangement in which the system is filled to as large a portion of the space as possible. This can contribute to sustainable development from the planet, people and profit perspectives. The three cases in this chapter illustrate that fill rate is an important packaging design direction for sustainable development. The cases consider sustainability impacts across the entire packaging system. Each case, though, has its own scope or approach that illustrates the impact of fill rate on the three pillars of sustainable development from an the organization, product or packaging perspective.

The first case demonstrates a well-functioning and well-designed packaging system, with a high volume of everyday products. It still has considerable improvement potential for developing more sustainable retail supply chains. The case compares the existing package with the suggested redesign. The second case compares two different packaging design solutions for ice cream and discusses their impact on sustainabile development. It demonstrates how effective the fill-rate direction can be in packaging design, but also the many trade-offs that need to be considered. The third real-life case evaluates the sustainability outcome from a decade-long implemention of a new unit load carrier throughout the global retailer IKEA's product range and supply network – from manufacturers via distribution centres to retail stores worldwide.

All three cases clearly demonstrate that there are significant sustainability implications related to the fill rate of packaging systems. That is why fill rate is an unquestionable direction in the packaging design compass for sustainable development.

The cases are:

- Detergent powder packaging: Less is more
- Ice cream packaging: Brick or elliptic ice cream?
- IKEA loading ledges: It's not rocket science, but it is about space.

Managing Packaging Design for Sustainable Development: A Compass for Strategic Directions, First Edition. Daniel Hellström and Annika Olsson.
© 2017 John Wiley & Sons, Ltd. Published 2017 by John Wiley & Sons, Ltd.

8.1   Detergent powder packaging: Less is more

Case by Daniel Hellström

One of the most basic products found in households worldwide is detergent. The latest detergent development is the introduction of new ingredients to further concentrate the product. As a result, many detergents have been reduced to a half or a third of their former volume. This means that each downsized package washes the same amount of clothes and linens as the bulky old package that it replaced. The environmental benefits of concentrated detergents are that they require less plastic for bottles, less corrugated cardboard for crating, and less fuel for the trucks that deliver them to stores. Retailers also benefit from this development. Reducing the volume allows a wider array of package designs and brands on store shelves.

In Sweden alone, approximately 40,000 tons of detergent are annually used in households, resulting in an annual detergent consumption of about 4.5 kg per person. Detergent is available in various forms, such as powder, liquid, gel, packaged as tablets and flakes. It is the detergent powder, though, that dominates the Swedish market with a share of approximately 80%. This is a market to be reckoned with for manufacturers who annually spend large sums on advertising their brands. Some of the brands found in Sweden are *Willys, Nopa, Skona, Lina, Neutral, X-tra, Ariel* and *Via*. All these brands use a similar primary package, a cardboard box.

Even though the introduction of concentrated detergent has resulted in downsized primary packages with environmental and economic improvements, the following question is posed: What additional packaging improvements can be made that would contribute to sustainable development? To explore this question, an evaluation of a packaging system for product detergent powder was carried out. The *Via* brand packaging system was chosen, since it is one of the leading powder detergent brands in Sweden. However, the solution is viable for other brands as well.

8.1.1   The detergent powder packaging system

The primary package for *Via* detergent powder comes in various sizes. The dimensions of the package we chose to evaluate were 180 × 140 × 62 mm and contains 900 g. This primary package we evaluated is a laminated cardboard box with a perforation on the upper part of one side that the consumer can press with his or her finger to create a pouring spout Figure 8.1). The secondary packaging consists of twelve primary packages enclosed in shrink film that is bottom sealed. There are no openings or handling devices for the secondary packaging.

A variety of different tertiary packaging is used for *Via* detergent powder. Some examples are EUR pallets (1200 × 800 mm), EUR half pallets (800 × 600 mm), display pallets, and roll containers of various dimensions. However, the EUR pallet is the one most commonly used and, consequently, the focus of this evaluation. Currently, a pallet contains 8 secondary packaging units per layer, 8 layers

Figure 8.1 *Via* 900 g primary packaging.

in height, which equals 576 primary packages per pallet (Figures 8.2 a–d). These are wrapped with stretch film, which forms the tertiary package.

8.1.2 The detergent powder supply chain

A supply chain for *Via* detergent powder is illustrated in Figure 8.3. In addition to the actors who transport products between the locations, the supply chain includes the following:

- producer in northern Italy;
- Unilever's distribution centre in southern Sweden (Åstorp);
- retail distribution centres;
- retail stores; and
- consumers.

Via detergent powder is produced by an Italian company situated in northern Italy, close to Milan. The *Via* brand owner, Unilever, hires different companies to transport the product from Italy to Sweden and the means of transportation vary over time. Sometimes it is by truck, and sometimes by a combination of truck/railroad where the container fits both on the truck and the train, referred to as "intermodal transport". The products arrive at Unilever Sweden, which is the national distribution centre for non-food products in southern Sweden. The

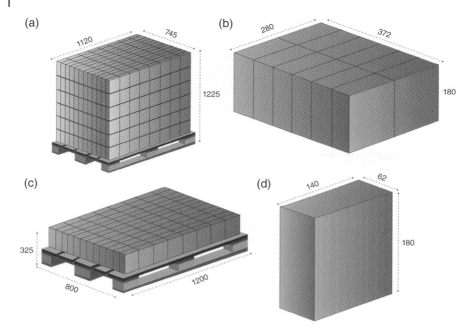

Figure 8.2 Illustration of the current *Via* detergent powder packaging system configuration: a) six layers of secondary packing units on a EUR pallet forming the tertiary package; b) eight secondary packaging units on a EUR pallet; c) one secondary packaging unit; and d) one primary packaging unit.

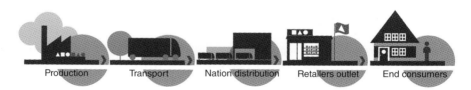

Production   Transport   Nation distribution   Retailers outlet   End consumers

Figure 8.3 A *Via* detergent powder supply chain.

products from Italy are unloaded and stored in the national distribution centre for further transportation to the retail distribution centres. Depending on the order quantity from the retail stores, the products are shipped from the retail distribution centres to the stores using different pallets and roll containers. Finally, the products are placed on the shelves of the retail stores.

8.1.3 Suggested packaging improvements

Insights were gained into the various requirements posed by the product, the packaging system and the supply chain actors after studying the interactions between the three. This resulted in the following two potential improvements in the primary package:

1) reducing its height; and
2) relocating and introducing a reclosable opening device.

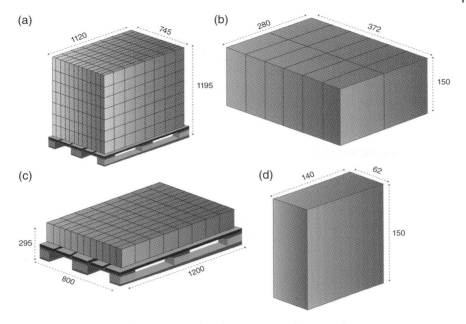

Figure 8.4 Illustration of the suggested *Via* detergent powder packaging system configuration: a) seven layers of secondary packing units on a EUR pallet forming the tertiary package; b) eight secondary packaging units on a EUR pallet; c) one secondary packaging unit; and d) one primary packaging unit.

The product fill rate of the primary packaging was considered low. Based on this, a suggestion was to reduce the package height by 30 mm. This resulted in one additional layer of products, which increased the total number of products by 17%, from 576 to 686 per pallet (Figures 8.4 a–d). The height can be reduced even more if needed. One trade-off when reducing height is a smaller shelf display surface. Another is that investments are most likely needed by the producer to improve the filling process (e.g. adding a shaker machine) since the powder needs to be made more compact.

Reducing the height of the primary packaging meant that the opening device needed to be relocated from the upper side to the top. Reducing the height without relocating the opening would have made the primary packaging difficult to open, but primarily it would have caused leakage due to the improved fill rate. A reclosable opening device was proposed in order to decrease potential leakage and enable an easier dosage/pouring process for the consumer. A prototype to show the principle of this improvement is shown in Figure 8.5. This type of opening solution is currently found in a variety of products on the Swedish market, such as salt.

8.1.4 Potential implications

It is clear that when evaluating the improvements made to the packaging system, we needed to see not only the direct positive results, but also indirect effects. The reduction of the primary package's height benefits the producer, whose costs

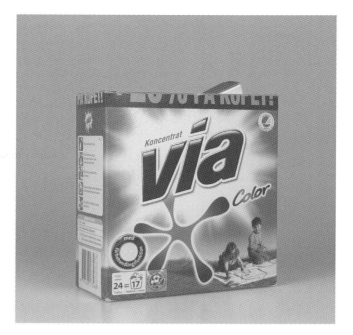

Figure 8.5 A prototype of a suggested opening device.

are reduced because less packaging material is needed for primary, secondary and tertiary packaging system components. It also results in better utilization of the pallet, which benefits the distribution centres by reducing product handling (fewer deliveries, less forklift movement) and the space needed in warehouses. Likewise, fewer transports are needed because of the elimination of empty space in the primary packaging. This reduces the total volume of empty space to be transported but increases the total volume of detergent powder that is transported. The reduced number of pallets to be used also reduces their return flow. One needs to keep in mind that one transport in every eleven is a return transport when using EUR pallets. Further down the supply chain, the retail store benefits from the improvement since less shelf space is needed, enabling the store to allow a wider array of products and brands on store shelves. For consumers, the new opening device can enable an easier dosage and decrease the potential of leakage. The major weaknesses of the improvements are the associated investment costs needed in the filling process at the producer and that the primary packaging will reduce the size and thereby the facing of the product display in the retail store.

8.1.5   Concluding remarks: Less is more

This case shows that a well-functioning and well-designed packaging system with a high volume of commonly used products, can be modified to develop a more sustainable design. More specifically, the case shows in detail that there are significant sustainability implications related to the fill rate of the packaging

Table 8.1 The impact (+ pros and – cons) on the three pillars of sustainability, of enhanced fill rate for the detergent powder packaging system.

	Planet	People	Profit
Producer	+ Less packaging material		+ Less packaging material
			+ Less handling
			– Less product display in store
			– Filling equipment investments
Transport	+ Fewer transports		+ Fewer transports
Distribution Centre	+ Less warehouse space needed		+ Less warehouse space needed
	+ Less forklift truck movement		+ Less handling
Distribution	+ Fewer deliveries		+ Fewer deliveries
Retail	+ Less shelf space needed	+ Easier shelf replenishment	+ Less shelf space needed
Consumer	+ Less product waste	+ Less packaging material to recycle	+ Less product waste
		+ Easier to carry	
		+ Easier to handle	
		+ Less space required	

system. For detergent powder, the suggested improvement of reducing the height and introducing a new reclosable opening device located on top of the primary package has the potential to influence sustainability across the whole supply chain. Some parts of the supply chain will benefit at the expense of the producer (Table 8.1). From planet and profit perspectives, the major benefit is the increase in fill rate of transport units, warehouses and retail shelves, while the main cost drawbacks are investments by producers and decreased product display at retail stores. The introduction of the suggested packaging improvements also has the potential to cut costs associated with packaging material and product handling. From a people perspective, the improvements make retail shelf replenishment easier for the employees and handling of the product easier for the consumers. The suggested opening device can result in less product waste, since it is resealable and helps the customer in product apportionment.

Case acknowledgements

Special thanks to Märtha Sjögren, Jenny Persson and Magnus Fagerlund for their contributions to the interviews and the prototype.

8.2   Ice cream packaging: Brick or elliptic shape?

Case by Daniel Hellström

Ice cream is an iconic summer treat enjoyed around the globe in many variations. With an annual consumption of approximately 12 litres per capita, Sweden ranks among the top five in ice cream consumption in the world. Two of the leading ice cream companies in Sweden manufacture 0.5 litre products that are sold through the three domestic grocery retailer giants in Sweden; they account for approximately 70% of total grocery market share in that country (Hultman and Elg, 2012). These two products have different packaging systems primarily due to marketing issues, such as brand differentiation and recognition. One of the products has a primary package shaped like a brick, while the other has an elliptic shape. In this case, these systems are compared in terms of their impact on the three pillars of sustainable development to gain insights into the fill rate direction of the packaging design compass. The two packaging systems are described in sections 8.21 and 8.22.

8.2.1   *GB Glace* brick packaging system

GB Glace is a subsidiary of Unilever, which is the world's biggest ice cream manufacturer. Almost half of all ice cream sold in Sweden is *GB Glace*. One of their best sellers is their 0.5 litre ice cream, sold in brick-shaped packages measuring $150 \times 90 \times 40$ mm (Figure 8.6). The shape has its origin from the 1950s when small freezers were introduced to most everyone's home. Even though the packaging solution has been technologically improved over the years, it is still shaped like a brick because it has a high emotional value for consumers. The packaging material is

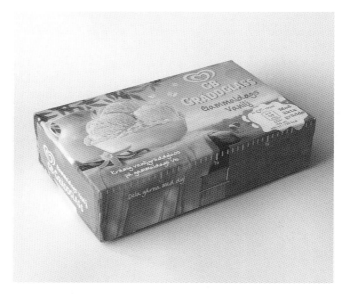

Figure 8.6 Primary packaging for *GB Glace* 0.5 litre ice cream.

polypropylene laminated carton. Approximately 12 million packages are produced and sold annually. The product price is approximately €1.2–1.4 to the consumer.

The secondary packaging contains ten primary packages that are shrink wrapped. It measures approximately 395 × 149 × 91 mm (Figure 8.7).

Eighteen layers with 15 secondary packaging units on each layer are placed on a pallet. In total, the pallet holds 270 secondary packaging units, which corresponds to 2,700 units of 0.5 litre ice cream per pallet (Figure 8.8). It is worth nothing that the total height of the tertiary packaging is 1,797 mm.

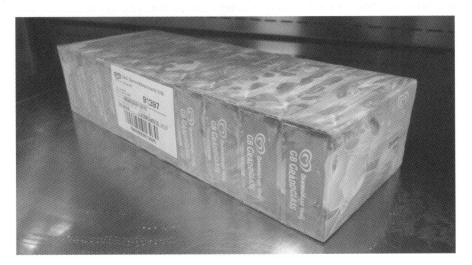

Figure 8.7 Secondary packaging for *GB Glace* 0.5 litre ice cream.

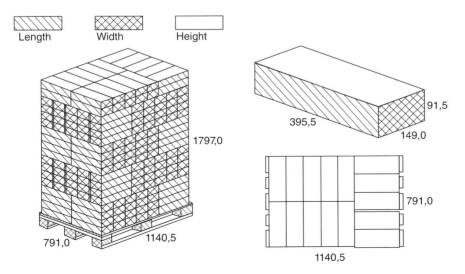

Figure 8.8 Illustration of the tertiary packaging system for *GB Glace* 0.5 litre ice cream. All measures are in millimetres.

Figure 8.9 Primary packaging for 0.5 litre *SIA Glass*.

8.2.2 *SIA Glass* elliptic packaging system

SIA Glass is a family business and its concept is to develop, manufacture and sell ice cream and other frozen products. One major product range for *SIA Glass* is ice cream in elliptic plastic (polypropylene) boxes with tamper-evident sealing. The product range has various flavours, sizes and qualities. The product line clearly distinguishes itself from other products on the market and the entire product range competes by means of a differentiation strategy. The primary packaging shape is important from a marketing perspective, and to enhance brand recognition, the elliptic shape is the same for all the products.

The 0.5 litre products (Figure 8.9) have an annual demand of approximately 5 million packages. The product fill rate is high, since the box has space for slightly over 0.5 litre. The shape of the lid facilitates stacking. At retailers, the product price to consumers is approximately €1.3–1.5.

The secondary packaging is a corrugated cardboard box containing eight primary packages, which are stacked in two layers (Figure 8.10). The design of the secondary packaging is open on two sides. This is primarily to save material and reduce costs, but the design also facilitates easy gripping and product recognition. The dimensions are approximately $295 \times 200 \times 150\,\text{mm}$.

Seven layers with 16 secondary packages (cardboard boxes) on each layer are stacked on a pallet. In total, the pallet holds 112 secondary packaging units, which corresponds to 896 units of 0.5 litre ice cream per pallet (Figure 8.11). The total height of the tertiary packaging pallet is 1,200 mm.

8.2.3 Supply chain descriptions

A key distribution feature of ice cream is that it must be kept frozen at all times to prevent the formation of ice crystals, which cause a grainy texture. It is best kept at or below $-18\,°\text{C}$. However, the distribution set-up can be different, as is the case when comparing *SIA Glass* and *GB Glace*.

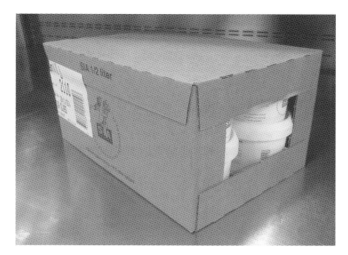

Figure 8.10 Secondary packaging for 0.5 litre *SIA Glass*.

Figure 8.11 Pallets of *SIA Glass* at a distribution centre.

SIA Glass ice cream is distributed through the retailers' regular cold chain, that is, via distribution centres (Figure 8.12). However, the replenishment at retail outlets is different compared to all other frozen products delivered from distribution centres. Once a week, manufacturer representatives visit the retail outlets to carry out product replenishment. This means that the retail staff moves the ice cream to a separate roll container, and replenishes the other frozen products that

Figure 8.12 *SIA Glass* supply chain.

Figure 8.13 *GB Glace* supply chain.

have been delivered. If the retailer receives more than one delivery of frozen products a week, which is nearly always the case, the retail staff handles the replenishment of the *SIA Glass* ice cream on those days.

In contrast to *SIA Glass*, *GB Glace* distributes its ice cream products directly to retail outlets via a nationwide distribution network dedicated to *GB Glace* products, made up of more than 30 depots (Figure 8.13). Sales representatives at the depots drive to retail outlets, check the amount of ice cream the store needs, go back to the truck and pick up the ice cream, and then replenish the freezer inside the store.

8.2.4 Comparing packaging solutions: A scenario

Because different supply chains pose different needs and requirements of packaging systems, we cannot simply compare packaging designs without considering the supply chain implications. To make a comparison between the two ice cream packaging solutions, we will analyse the following scenario: What would be the sustainability impacts if *SIA Glass* ice cream went from an elliptic to a brick packaging system? In this scenario, the *SIA Glass* ice cream is still distributed through retailers' cold chain as described in section 8.2.3.

Also, in such a scenario, the "new" *SIA Glass* brick packaging system can have a maximum pallet height of 1,200 mm to comply with the retailers' distribution centre directive. The pallet would hold 11 layers containing 1,650 primary packages of 0.5 litre ice cream bricks. Compared to the elliptic solution, which contains 896 primary packages, the "new" brick solution has 84% more ice cream on each pallet.

With 84% more ice cream on each pallet, the handling and transport activities from the producer to the distribution centre would be reduced by approximately 46%. For *SIA Glass*, with an annual demand of 5 million primary packages, the number of tertiary packages would be reduced from approximately 5,580 to 3,030. This corresponds to 77 fewer frozen transports using EU standard 16.5 m trailers. For *GB Glace*, with an annual demand of 12 million primary packages, going from a brick to an elliptic solution would result in an increase from 7,270 to 13,390 units of tertiary packages. This corresponds to 185 more frozen transports using 16.5 m trailers. An additional benefit of the brick packaging system is less handling of incoming packaging material at the producer. Thus, there is an evident impact on profit and planet, since transport costs are approximately €10 per 10 km and the fuel consumption is 3.5–3.8 litre per 10 km, which equals approximately 2.6 kg CO_2e per litre fuel.

An important impact of improving fill rate for frozen products is that less temperature-controlled space is needed throughout the supply chain. This is something that retailers, for example, have requested, because empty space in the freezer increases the energy consumption needed for cooling, as a full freezer retains cold better and the mass of cold items will enable it to recover more quickly. Less space needed in frozen warehouses results in decreased energy consumption, approximately 50–60 kWh/m^3, which in turn reduces costs. The daily cost of a single pallet position at a frozen warehouse is approximately €0.3. Assuming 2,550 less units with an average of 60 days in the warehouse, this results in an annual cost reduction of €45,900. Fewer frozen transports also decrease energy consumption and reduce the carbon footprint and its associated costs.

In the replenishment process, the retail staff often prefer the corrugated cardboard box solution, since it is easier to open and provides a good grip, while the *GB Glace* shrink wrap solution makes it humid, slippery and needs to be opened with a knife. The corrugated cardboard box solution also has better insulation properties, which protect the ice cream from potential heat.

As is true with consumers, one design does not fit all. Hence, end consumers have different opinions about the two primary packaging solutions. The elliptic one has an important second use. It is one of the most popular lunch boxes in Sweden and can be used in a microwave. It is resealable which means that it protects the contents from any contamination after it has been opened. The brick solution has a fill rate advantage. End consumers also like the easy portioning and serving of the ice cream that the brick solution affords: You simply cut a piece off the brick with a knife. What remains when you put it back in the freezer does not utilize any unnecessary space and it is easy for the end consumers to identify the amount of ice cream left without opening the primary package.

8.2.5   Concluding remarks: Brick or elliptic shape?

The case shows that there are significant sustainability implications related to the fill rate direction of the packaging design compass, in the choice of packaging shape. Table 8.2 shows the sustainability impacts of going from an elliptic to a brick solution. This illustrates that a simple packaging decision concerning the shape of a primary packaging has far-reaching consequences on sustainability, especially for frozen (temperature controlled) high volume products such as ice cream.

Even though these two products are sold through the same retail channels, they have different supply chain set-ups. Without considering the sustainability implications of the supply chain set-up itself, and strictly comparing the packaging systems, there are several impacts from a planet perspective. Less space needed along the supply chain, from producer to consumer, means more efficient

Table 8.2 The impact (+ pros and – cons), on the three pillars of sustainability, of going from an elliptic to a brick packaging system.

	Planet	People	Profit
Producer	+ Less packaging material		+ Less packaging material
	+ Less handling of incoming pallets		+ Less handling of incoming pallets
	+ More use of renewable resources for primary packaging		– Decreased perceived product quality
	– Less use of renewable resources for secondary packaging		– Less product display in store
			– Decreased sales
Transport	+ Fewer frozen transports		+ Fewer frozen transports
Distribution Centre	+ Less cold warehouse space needed		+ Less cold warehouse space needed
	+ Less cold storage energy needed		+ Less cold storage energy needed
	+ Less forklift truck movement		+ Less handling
Distribution	+ Fewer frozen deliveries		+ Fewer frozen deliveries
Retail	+ Less freezer space needed	– More difficult to open secondary packaging	+ Less freezer space needed
	+ Less freezer energy needed	– More difficult to replenish freezer	+ Less freezer energy needed
Consumer	+ Less freezer space and energy required	+ More easy to portion	+ Less freezer space required
	+ Less packaging material to recycle		+ Less freezer energy needed
	– No "second use" of primary packaging		– No "second use" of primary packaging

use of warehouse and retail space and fewer transports. This decreases the environmental impact by decreasing the energy consumption needed for cooling and through more efficient land utilization. Less packaging material and greater use of renewable resources are aspects that also improve sustainability from a planet perspective.

From a people perspective, the main impact in going from the elliptic to the brick packaging solution is in the replenishment process at retailers, where the retail staff prefer the *SIA Glass* corrugated cardboard box solution for its secondary packaging, since it is easier to open and provides a good grip. The *GB Glace* shrink wrap solution, on the other hand, results in a moist, slippery surface and needs to be opened with a knife.

From a profit perspective, all supply chain actors benefit from an improved fill rate in the form of less transport, less handling and lower energy costs. However, for *SIA Glass* there are marketing consequences such as decreased perceived product quality and less product display in stores. These are vital to consider because they can eventually result in decreased sales. Hence, for *SIA Glass* changing shape is not an option, while *GB Glace* is able to take advantage of the brick solution both from a logistics and a marketing perspective. This case shows that there is not always, but can be, a trade-off between the fill rate direction and the information and communication direction of the packaging design compass. For both ice cream products in this case, the shape of the primary packaging is a key aspect in branding.

In conclusion, there are pros and cons for these two packaging systems. From a strict fill rate point of view, the brick system is better. But from a sustainability perspective, neither can be determined as better or worse. It depends on the supply chain set-up and how the pros and cons are balanced.

Case acknowledgements

Special thanks to Dominika Hagborg (Marketing Manager Ice Cream) at Unilever, Lars Hillertz (Chief Technical Officer) at *SIA Glass*, and Sampo Heikkinen (Product & System Specialist) at ICA Sverige for product and packaging descriptions.

8.3   IKEA loading ledges: It's not rocket science, but it is about space

Case by Daniel Hellström

The pallet is arguably as fundamental to globalization as the container (Fabbe-Costes et al., 2006; Kearny, 1997; Levinson, 2006). Using pallets for loading and unloading results in lower costs for handling and storage, and faster material movement. In many organizations, the use of pallets is hardly even questioned. But not everyone thinks that the pallet is the unit load carrier that should be used. The global retailer IKEA, a company well known for its fixation on logistics and packaging, has abandoned wooden pallets in favour of other unit load carriers, such as loading ledges – an innovation that enable variations in unit load dimensions. The introduction of loading ledges has had a profound impact on IKEA's global supply chain. That is why it is of interest to investigate the impact of this unit load carrier on the three pillars of sustainable development.[1]

---

1 Parts of this case were published in Emerald's *International Journal of Retail and Distribution Management*, (2011) by Hellström and Nilsson, "Logistics-driven packaging innovation: a case study at IKEA".

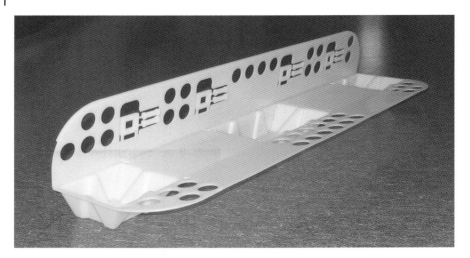

Figure 8.14 A loading ledge.

A loading ledge is one component made of recyclable polypropylene plastic that weighs about 370 g (Figure 8.14). Several of them are used to construct a unit load carrier. Depending on the shape of the unit load, the loading ledges are placed in different positions beneath the products. The loading ledges are then strapped to the products and the unit load is wrapped in stretch film in order to hold it together and stabilize it. Loading ledges are furthermore stackable and can handle up to 5,000 kg of static pressure.

If you have products that are packed in relatively small units compared to their unit load dimensions, you need a supporting platform when using loading ledges. Particle board works well as a base for these kinds of products. This also increases the stability of the unit load and the protection it offers (Figure 8.15).

The main difference between using loading ledges and a traditional pallet is that loading ledges allow for varying size and design of products. Traditional unit load carriers, like pallets, have fixed dimensions which means that products and packaging are designed and produced accordingly. With loading ledges you are able to adjust the unit load dimensions to specific needs and requirements of the product. Instead of the product dimensions being modified to fit the load carrier, the load carrier is adjusted to fit the products. Another difference is that loading ledges do not constitute a self-supporting unit load carrier. The load-bearing support of the unit load comes from the products instead. For products that are non-supporting, it is essential that the loading ledge packaging system provides a supporting platform. Yet another difference is that loading ledges only add 45 mm to the height of a unit load, while a wooden pallet adds 145 mm.

Material properties also generate differences in unit load carrier characteristics. Plastic as used for loading ledges, is a relatively expensive material, but can be shipped all over the world without border restrictions or extra treatment considerations, which is not the case with packaging material made of wood. Wood has to comply with the regulatory requirements of the *International Standard for Phytosanitary Measures* to reduce the risk of introducing and/or spreading

Figure 8.15 How to use particle boards as supporting platforms with loading ledges.

quarantine pests associated with wood packaging material (FAO, 2006). Corrugated cardboard is obtainable all over the world but is sensitive to humidity, which negatively influences the load-bearing qualities and stability of corrugated cardboard pallets.

8.3.1 Implementation – from 2001 to 2010

The loading ledge as a unit load carrier has been implemented throughout IKEA's product range and global supply chain network, from manufactures via distribution centres to retail stores worldwide. It was first introduced at IKEA in 2001. In 2006 the loading ledge was being used for approximately 10% of the total inbound flow. These numbers were continuously rising and by 2010 its use was up to 30%.

The introduction of the loading ledge was interconnected with IKEA's product sourcing strategy. In 2001 IKEA was sourcing 66% from Europe, 30% from Asia and 4% from North America. In a five-year period, IKEA expected to increase the sourcing from Asia by 15–20%, generating new needs and requirements on the distribution process. One need was to address the lack of standardized unit load carriers in Asia. Thus, implementing the loading ledge was one measure to reach these figures. By 2010 IKEA was sourcing 62% from Europe, 34% from Asia and 4% from North America. However, during 2001 until 2010, IKEA has more than doubled its total volume of products sold. According to IKEA, the implementation of the loading ledge has been a key element in its global sourcing strategy.

8.3.2 Supply chain impact

The introduction of loading ledges has consequences throughout supply chains. The consequences for manufacturers, transporters, distribution centres, retail stores, and the recycling and reuse system are described in sections 8.3.2.1–8.3.2.5.

8.3.2.1   Manufacturers

For manufacturers that supply IKEA, the shift from traditional unit load carriers to loading ledges has meant replacing the existing palletizing equipment or investing in new items. Automated palletizing machines (using loading ledges instead of pallets) have been implemented in the existing production lines on the premises of some manufacturers. Other manufacturers prefer manual palletizing, given the low cost of labour, where a packaging fixture is used to strap the loading ledges to the products. In automated production lines, one result of using loading ledges instead of wooden pallets is that there are fewer stops in production. Wooden pallets are often rejected due to their poor quality and this halts production.

A manufacturer often serves different markets and each has its unit load carrier preferences. The European market prefers the EUR pallet, which is not an option for other markets, such as those of North America and Southeast Asia. Thus, the manufacturer needs to have different types of unit load carriers for different markets. Introducing loading ledges has made it possible for some manufacturers to serve all markets with one type of carrier, thereby simplifying their production and planning, and reducing inventory.

8.3.2.2   Transport

Using loading ledges provides an opportunity to increase the fill rate of transport units (railway cars, containers and trailers). Traditional unit load carriers sometimes limit the fill rate of transport units because they are not always compatible with the dimensions of the products, and this creates empty spaces between the unit loads in transport units. Loading ledges allow unit load dimensions to be adjusted to the products, thereby eliminating empty spaces.

The increase in fill rate depends on the design of the product and on the mode of transport. In shipments that reach the regulated weight limit, the fill rate can be increased by as much as 3% when wooden pallets are replaced with loading ledges, since the loading ledges are lighter. Adjusting the unit load carrier to the dimensions of the product has also resulted in less damage to products. Fewer empty spaces between the unit loads have resulted in less movement within the transport unit, which in turn has decreased the risk of damaging goods. In order to illustrate the impact of loading ledges on transport, an example will be described and evaluated.

The "600-mm products" are a high-volume group at IKEA. These include wardrobes, cabinets and bookcases which are 600 mm wide and up to 2.45 m long. The long pallet was traditionally used as the load carrier for these products. However, this left empty spaces between the unit loads in transport units, resulting in poor utilization of the transport volume and risking product damage due to movement in transport units. Instead, by using four loading ledges strapped to products to form the unit loads, the empty spaces between were eliminated. Figure 8.16 illustrates a total fill rate increase of 44%. The average increase in fill rate, however, is about 26% for 600–mm products. The filling increase is due to the elimination of empty spaces between units as well as the utilization of the space previously taken up by pallets.

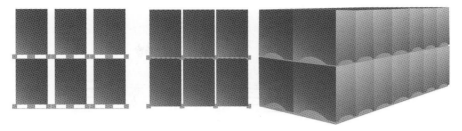

Figure 8.16 Illustration of a 44% fill rate increase of a container.

In 2006 IKEA had an annual volume of 3 million m^3 of 600-mm products. Using loading ledges on all these products, and assuming a 20% average increase in fill rate, this would result in 600,000 m^3 less transport volume per year. This is equivalent 10,000 12-m trailers.

8.3.2.3 Distribution centres

The introduction of loading ledges had an initial detrimental effect on operations at distribution centres. The existing warehouse and material-handling systems, such as storage rack configurations, forklifts and conveyors were designed for standardized wooden pallets and could not accommodate the varying dimensions of unit loads caused by the loading ledges. Thus, investments in adjusting the infrastructure were needed. In Europe alone, more than 500,000 new metal shelves had to be installed at distribution centres and retail stores. However, adjusting all the infrastructure (e.g. automated warehouse systems) at distribution centres to loading ledges is not economically feasible. So in order to handle some loading ledge units at distribution centres, IKEA currently straps them to wooden pallets. Doing so is an additional, time-consuming activity. Automated strapping equipment was introduced at the distribution centres to reduce extra handling time. In addition, loading ledge units on wooden pallets occupy more storage space and this has led to reduced fill rates at the warehouses. One example of where loading ledge units are not strapped to wooden pallets is when they are stored in block storage systems, where they increase the utilization of storage space by eliminating the empty spaces between units.

The strapping of loading ledge units to wooden pallets is the same handling procedure as one uses when dealing with corrugated cardboard pallets, but requires less handling time. A majority of the non-European material flows use corrugated cardboard pallets that are transported by sea. This frequently means that the pallets collapse during transport due to humid transport conditions. Loading ledges offer better protection from humidity and forklift handling, and this facilitates a more efficient unloading process at the distribution centres.

8.3.2.4 Retail stores

By 2006, loading ledges had not yet had a significant impact on retail because stores did not have the necessary material handling equipment to receive or handle them. The existing IKEA stores had been built and designed, just like the distribution centres, to handle wooden pallets. However, in the process of increasing the amount of direct deliveries, stores have developed the ability to handle an assortment of load carriers, such as loading ledges, and not just wooden pallets. To accommodate different unit load carriers in its stores, IKEA has been working intensively with its material handling equipment suppliers to modify and improve the equipment so that it can be used to handle different unit load carriers.

A great potential in using loading ledges at stores is to improve the display function in the sales area. A majority of incoming goods are directly transported to the sales area. Wooden pallets in sales displays have sometimes been found to be less appealing to customers. IKEA has developed various packaging sale solutions where loading ledges are integrated into trays and display packaging. This improves the display function. According to IKEA, this improvement potential has not yet been fully implemented or evaluated.

8.3.2.5 Recycling and reuse system

Loading ledges can be reused and recycled. In recycling, they are ground down into plastic pellets and then used as raw material for new loading ledges. Compared to the wooden pallet return system (which costs IKEA about €30–35 million annually), the use of loading ledges drastically reduces the number of return transports and the need for storage space, since they take up much less space. An ordinary trailer has the capacity to carry approximately 50 unit loads (using for example the EUR pallet). The same trailer has the capacity of 500 empty EUR pallets or more than 34,000 empty loading ledges units. This means that one transport in every eleven is a return transport when using EUR pallets, compared to one in every two hundred when loading ledges are reused. Figure 8.17 illustrates the stackability and a unit load consisting of 500 loading ledges.

8.3.3 Concluding remarks: It's not rocket science, but it is about space

The case shows that there are significant sustainability implications related to the fill rate. For IKEA, the introduction of loading ledges has influenced the sustainability of the whole supply chain with some parts benefiting at the expense of others (Table 8.3).

From a planet perspective, the use of loading ledges instead of wooden pallets has shown that there is a potential to reduce the negative environmental impact of transport. A life-cycle assessment comparing the environmental performance of using loading ledges to EUR pallets indicated that the differences in their environmental impacts were irrelevant (Strömberg et al., 2003). They concluded that the mode of transport is a much more important aspect to consider than type of unit load carrier (ibid.). However, the life-cycle assessment only measured the environmental transport impact per ton-kms. This means that the capability to increase the fill rate of transport units is not included, which reduces the amount of vehicle movement, that is, vehicle-kms. Consequently, loading ledges themselves may not make a smaller planet footprint than the EUR pallet, but they enable higher fill rates of transport units, which in turn reduces the negative environmental impact of transport.

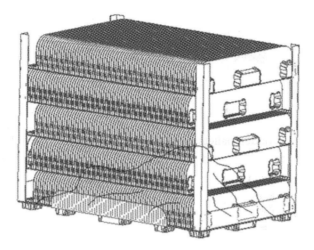

Figure 8.17 Illustration of a unit load consisting of 500 loading ledges.

From a people perspective, the loading ledges enable IKEA to meet new needs and requirements from different markets. This assists IKEA in sourcing and reaching new markets such as in developing countries. For example, a third of its material flow originates from markets where the use of wooden pallets is not a viable choice. Here, loading ledges represent an alternative option to corrugated cardboard pallets. Loading ledges are an additional option in the choice of load carriers, which enables the company to give greater consideration to differences in infrastructure and equipment between markets. Introducing loading ledges has also enabled more freedom in product and packaging design. Instead of being constrained by the dimensions of the load carrier, the creativity of product designers and packaging engineers who use loading ledges can be guided more towards sustainable development.

From a cost perspective, the major benefit is the increase in fill rate of transport units, while the main drawback is additional time-consuming activities at distribution centres. However, estimates from IKEA indicate that the current annual decrease of transport costs (more than €2 million) is more than ten times greater than the cost of additional handling at distribution centres. Using loading ledges has also cut costs associated with return handling and decreased the rate of damage to goods, pointing out synergies with the product protection direction of the packaging design compass.

8.3.4 Epilogue

In spite of this massive implementation during the first decade of this millennium and its many faceted benefits, the loading ledge was in the beginning of 2016 just being used for approximately 10% of the total inbound flow. The reason is that by 2020, all plastics used in IKEA must be either recycled or made from renewable bio-based raw materials. Not to use any newly fossil plastics is one step in the company's sustainability strategy. With this ambition the loading ledge has become a "non-cost-efficient" solution, since recycled plastic material

Table 8.3 The impact (+ pros and − cons) on the three pillars of sustainability, of introducing loading ledges throughout IKEA's product range and global supply chain network.

	Planet	People	Profit
Producer	+ Less warehouse space needed	+ Sourcing and reaching new markets	+ Increased production efficiency
		+ More freedom in product design	− Investment in equipment
Transport	+ Fewer transports		+ Fewer transports
	+ Fewer return transports		+ Fewer return transports
	+ Less product damage		+ Less product damage
Distribution Centre	+ Less warehouse space needed		− Investment in equipment
			+ Less product damage
Distribution	+ Fewer deliveries		+ Fewer deliveries
	+ Fewer return transports		+ Fewer return transports
	+ Less product damage		+ Less product damage
Retail		+ Less packaging material to handle	− Investment in equipment
		+ Easier carrying	+ Improved display
		+ Easier handling	

is often even more expensive than virgin material. IKEA is therefore still committed to, and are heavily investing in, developing alternative unit load solutions such as corrugated cardboard pallets, which according to IKEA often is a better solution from a total cost perspective.

Case acknowledgements

Special thanks to the packaging development team at IKEA of Sweden. Foremost gratitude to Allan Dickner (Deputy Packaging Manager) and Mikael Lindmark (Packaging Developer) for sharing the roller-coaster journey of implementing Loading Ledges and their cutting-edge knowledge of Packaging Logistics in global supply chains.

References

Fabbe-Costes N., Jahre M. and Rouquet A. (2006), Interacting standards: A basic element in logistics networks. *International Journal of Physical Distribution & Logistics Management*, 36(2), 93–111.

Food and Agriculture Organization (2006), *International standards for phytosanitary measures*: (http://www.fao.org/docrep/009/a0450e/a0450e00.htm)

Hellström D. and Nilsson F. (2011), Logistics-driven packaging innovation: A case study at IKEA. *International Journal of Retail and Distribution Management*, 39(9), 638–657.

Hultman J. and Elg U. (2012), Country Report Sweden. *European Retail Research*, 26(2), 151–166.

Kearny AT. (1997), *The Efficient Unit Loads Report*. ECR Europe, Brussels.

Levinson M. (2006), *The Box: How the Shipping Container Made the World Smaller and the World Economy Bigger*. Princeton University Press, Princeton, NJ.

Strömberg K., Fröling M. and Berg H. (2003), *Comparative LCA of Loading Ledge and Wooden Pallet*. CIT Ekologik AB, Gothenburg.

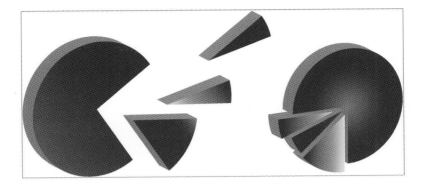

9   Apportionment

In packaging design, apportionment is often based on many years of tradition, especially when it comes to numbers, weights and volumes of products in packaging. In this section we present three cases that explore the apportionment direction of the compass. The cases describe and evaluate apportionment effects on both primary and secondary packaging in relation to sales, product losses and the physical handling in supply chains.

The first is a real-life case that focuses on a sensitive product – fresh smoked salmon – and the effects two existing apportionment alternatives have, particularly on product waste for the consumers. The second case looks into toothpaste, an everyday consumer product. A suggested different apportionment of the number of primary packages in the shelf-ready secondary package is compared and evaluated. The third case discusses the apportionment of the secondary packaging of wine and liquor bottles. The traditional 12 bottles per secondary packaging is challenged and new apportionment alternatives are compared and evaluated.

The three cases are:

- Apportion for less product waste;
- Appropriate numbers in shelf-ready packaging;
- The quantity of bottles boxes.

9.1   Apportion for less product waste

Case by Fredrik Nilsson and Annika Olsson

Fish, meat and chicken are all food products that are relatively expensive and have high environmental impacts. These types of products are sensitive to handling and distribution in the supply chain, especially when they are distributed fresh. That is why it is essential to have reliable packaging solutions for cold food chains, solutions that protect the food from contamination, spoilage, damage and that maintain freshness.

The environmental impact from meat and fish are very high compared to vegetation based prime produce. This can be shown by comparing the environmental impact from the primary production of fish with the one of vegetables.

For example, the environmental impact of salmon filé itself is 6.6 kg CO_2e/kg (Buchspies et al., 2011; Nathan et al., 2007). This can be compared to frozen peas which have an impact of 0.33 CO_{2e}/kg (Figure 9.1). The total impact of salmon filé including primary production, secondary production, packaging, storing and transportation (i.e. logistics) is 7.5 kg CO_{2e}/kg, which means that 88% of the total environmental impact is due to primary production. When it comes to peas, the total impact is 0.53 CO_{2e}/kg, which means that 62% is derived from primary production (Nilsson and Lindberg, 2011). Other studies provide similar figures (emission per kg of fish filé ranging between 3.7 and 6.6 kg CO_{2e}[1]).

Consequently, every aspect that can have an impact on the minimization of losses of produced food products will represent effective ways of lowering the total environmental impact. In the case we present here, we examine the apportionment of salmon and show the potential this can have on lowering the loss of food.

This case examines how salmon, and in particular a product sold in Sweden called "gravlax", is apportioned. Gravlax consists of salmon filés that have been cured by being marinated in salt, sugar, dill and other spices. The focus here is on a comparison between the 400 g and the 100 g primary packages that are illustrated in Figure 9.2. This type of salmon is available in most of the grocery stores in Sweden. The product requires barriers for protection against air, smell, taste and microorganisms. Since the cured salmon is perishable, it also has to be kept cold during transport, storage and distribution.

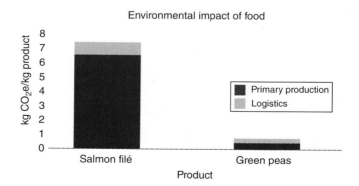

Figure 9.1 Environmental impacts of salmon vs. green peas (all measures include growth and fishing/harvesting, which are referred to as "primary production").

---

1 http://www.ecotrust.org/lca/

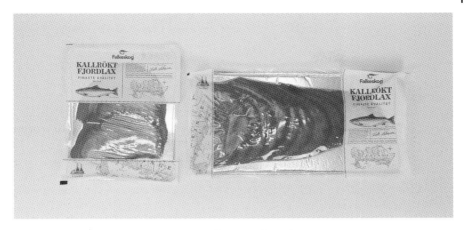

Figure 9.2 The 100 g and 400 g salmon packages evaluated.

The supply chain for salmon products starts at a fish producer located in Sweden. Based on the incoming fish from a number of suppliers, they process and pack the salmon (as well as other fish categories) and sell them to wholesalers, who in turn sell and distribute the fish to retail stores.

9.1.1 The salmon packaging system

The fish needs to be vacuum packed to ensure that it stays fresh, and the primary packaging needs to be cold stored during transportation. One requirement from the National Food Agency of Sweden is that the packaging has to be cooled down to below 0 °C before the fish is packed. Like most food products, the packaging is required to contain information about ingredients, the origins of the fish, expiration date and where it has been produced.

The primary packaging, illustrated in Figure 9.2, is made of plastic film which is joined together with a plate of laminated cardboard. The cured salmon is placed in the packaging and vacuum packed.

The secondary packaging is made of corrugated cardboard, holding up to twenty 100 g packages or five 400 g packages. The company only uses one size of the secondary packaging, the corrugated cardboard box. The corrugated cardboard box is produced as one piece and was chosen because the material can be recycled. It is easily sealed by simply putting one side of the lid in and under the other; no glue or tape is needed, as illustrated in Figure 9.3. The measurements (24 × 40 × 10 cm) are set to make a good fit on the pallet and according to the quantities required for most of the retailers per delivery.

The tertiary packaging can vary due to the different ways the packaging is handled throughout the supply chain. A EUR pallet is filled with secondary packaging at the manufacturer. At the most, up to 100 secondary packages can fit on one pallet. This limitation is necessary because the cardboard boxes would otherwise collapse if too many are stacked on top of each other. The pallet is wrapped with plastic film and transported to the wholesaler. The boxes are

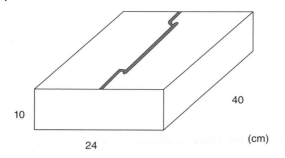

Figure 9.3 The secondary packaging used for salmon.

40

10

24

(cm)

picked from the pallets at the wholesaler and put in roll cages in the amounts the retailers have requested.

9.1.2 The impact of primary package apportionment

From a demand or consumer perspective, apportionment is an intriguing issue in relation to sustainable development. A larger primary package is economically beneficial for the consumer because the price per/kg for packages containing 400 g of salmon is €26.30/kg, while the price for 100 g is €35.40/kg. Hence, buying three 100 g packages (3* €3.54 = €10,62) instead of one 400 g package (€10.62) is a poor bargain because in the second package, you get an extra 100 g for pretty much the same price. At the same time studies show that people do not eat fish that has been opened once at home and then put back in the fridge. Instead people throw it away and buy new products. Assuming that 80 g of salmon is the average that people throw away from each 400 g package; with the environmental impact of salmon being approximately 7.5 kg CO_2e per kg salmon – this would mean that for 50,000 packages of salmon (0.4*50,000 = 20 tons salmon), 4 tons are thrown away resulting in an environmental impact of (4*7.5) 30 tons CO_2e. So, due to the price setting related to apportionment, it is cost beneficial for the consumer to throw those 80 g away because what he or she discards does not cost any extra. But from an environmental point of view this over-consumption, due to large volumes at low prices, has devastating effects.

Apart for the consequences in the consumer part of the supply chain, the apportionment of packaging also has consequences for the other supply chain actors. Some are related to standards and logistical requirements (pallets and information structures, for example) and others more easily adaptable (secondary packaging size and price of products, for example).

A number of improvement areas for the case of salmon packaging were found based on input from the supply chain actors' requirements. One of these was about the dimensions of the secondary packaging in relation to the quantity of primary packaging requested by retailers. When considering the quantities of the existing primary packaging, we found that the retailers' requirements would result in only 77% of the secondary packaging volume on average being utilized. Thus, a smaller secondary package would increase the fill rate in storage and transport and provide the retailers with less packaging material to handle.

This modification of the packaging system would mean that 47% more of the products could be placed on each pallet and, in turn, increase the area and volume efficiency in secondary and tertiary packaging. As a result, 31% fewer pallets would have to be handled and delivered to the wholesaler each year. This would free up space in the trucks or mean even fewer runs in total. The improvement would have both financial and environmental impacts, including decreased transportation costs, less pollution and reduced use of material use in secondary packaging. A risk related to the new solution is that the transport schedule would still be the same and this means that the truck would be less volume efficient. Still, less material and more efficient packaging will have more positive than negative effects. There may well be other potential trade-offs related to the investments needed in the packaging process.

9.1.3 Concluding remarks: Apportion for less product waste

While the pricing/apportionment situation is not unique in this case, most products can be bought in larger quantities based on the economic reasoning of lower price/kg. This view on apportionment has both environmental and social

Table 9.1 The impact (+ pros and – cons) on the three pillars of sustainability, from having 100 g packages for salmon filé, and more efficient secondary packaging.

	Planet	People	Profit
Producer	+ Less secondary packaging material		+ Lower cost for secondary packaging
	– More primary packaging material per product		– Higher cost for primary packaging
Transport	+ Less empty space being distributed	+ Less transports	+ Higher resource utilization
			+ Decreased transportation costs
	+ Decreased transportation costs		
Distribution Centre	+ Less empty space in storage		+ Less packaging material to handle
			+ More products per delivery
Distribution	+ Less empty space being distributed	+ More efficient handling	
Retail	+ Increased fill rate in secondary packages	+ Less secondary packaging material to handle	+ Less packages to handle
			+ Higher profit from smaller packages
Consumer	+ Less product wasted	+ Better quality per serving	– More expensive per kilo

implications for most of the actors involved. From a planet perspective, there are direct environmental consequences for every uneaten piece of salmon or any other product. From a people perspective, there are indirect social consequences in terms of the extra handling employees and companies have to carry out with smaller packages and thus larger package quantities (Table 9.1).

Case acknowledgements

Special thanks to Oskar Callenås, Johanna Olin and Christian Lycke Bladh for the initial work carried out.

9.2   Appropriate numbers in shelf-ready packaging

Case by Fredrik Nilsson

Toothpaste is a common product that is primarily sold in plastic tubes with either an extra carton box as primary packaging or as it is in a secondary, often shelf-ready package. When packed, the toothpaste is easy to handle and not very sensitive to normal temperature changes or rough handling. In this case, the ACTA brand for toothpaste and its supply chain were chosen, starting with the brand owner Cederroth's production in Holland, shipped to the Cederroth distribution centre in Sweden and then distributed to a number of pharmacies in Sweden.

9.2.1   The toothpaste packaging system

The ACTA toothpaste was the first in Sweden with the Nordic Swan Ecolabel and is produced without preservatives.[2] It comes in a plastic tube of 125 ml (height 195 mm, diameter 30 mm) and has been available for purchase at pharmacies since October 2009. Since the deregulation of the pharmaceutical market in the same year, it is also available at other retail outlets.

   The secondary packaging is a cardboard box ($110 \times 140 \times 200$ mm) in which 12 tubes are packed standing upright (Figure 9.4).

   On the tertiary level, depending on the supply chain, three different packaging alternatives are used: plastic containers for pharmacies; EU pallets with shrink wrap (often five layers of 56 boxes with 3,360 tubes); and/or roll containers for wholesalers and retail outlets.

9.2.2   The toothpaste supply chain

ACTA toothpaste is produced in Holland at the Cederroth facility. The annual volume is about 1.6 million tubes/year for the Swedish market. Using a third-party logistics provider, Cederroth ships their products for the Swedish market

---

2   Svanen, 2011 (www.svanen.se)

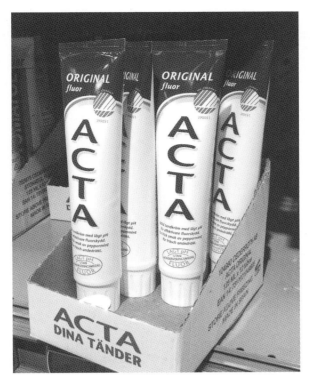

Figure 9.4 How the tubes are stacked in the shelf-ready secondary packaging.

on full pallets to their warehouse in Falun. Twice a week a mixed or full pallet, depending on the reorder point, is delivered to the wholesalers of pharmaceutical products in Sweden. There the different products are picked manually and placed in plastic boxes for the retail pharmacies, often mixed with other products to be delivered. Mixed boxes enable the wholesalers to maintain a high frequency of delivery, every 24 hour, even if the turnaround time for each product may be slow. The plastic boxes are delivered to local pharmacies where the toothpaste tubes are put up for sale on the shelves in their secondary package.

9.2.3   Reapportionment for easier handling and improved fill rates

It became clear in the evaluation of the packaging system that there was a lot of empty space in each secondary package due to the shape of the tubes (Figure 9.4).

By rethinking the current apportionment of 12 tubes per secondary package and instead filling the empty space with more products, this would result in an almost doubled quantity in the secondary package and more products per order (i.e. fewer orders and less handling). When examining it further, the proposed change meant a slight increase in the height of the boxes in order to fill the space between the tubes with another row of tubes placed upside down (Figure 9.5).

Because the secondary packaging is used to both display the product (shelf-ready solution) and facilitate the transportation, an insert was needed to stabilize

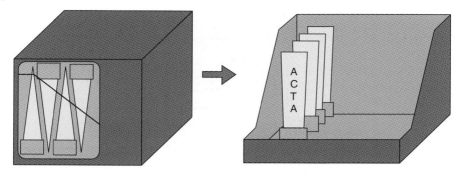

Figure 9.5 The proposed stacking of tubes in the secondary packaging.

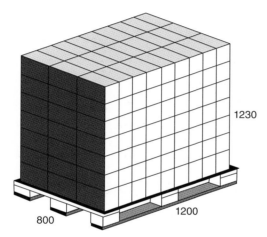

Figure 9.6 A pallet of secondary packaging containing 24 tubes as the new apportionment.

the upper tubes so that they could be removed in one unit of 12 instead of one by one if no insert was made. By adding a perforation to the box (the red line in Figure 9.5), it can be split in half – holding 12 tubes per half and in that way be displayed directly. With the addition of an insert, the shelf-ready packaging properties can also be improved because the tubes will stand more stable in the tray than before.

The insert material will be beneficial to the pharmacies, as it will increase the stability of the tubes when displayed on the shelf. This means a better and more accurate display of the product. For the consumer, the use of the insert material may be inconvenient, as it will make it more difficult to remove the tubes from the secondary package in which they are displayed at the pharmacies.

The effects of this change in apportionment would require new dimensions of the secondary package: 125 × 155 × 250 mm in order to fit both the inserts and the double stacking. The new solution would fit 4,536 tubes of toothpaste per pallet (Figure 9.6). This would result in a 35% increase in fill rate. In total this would mean a reduction of approximately 120 pallets to be transported per year to and within Sweden, also meaning less handling of packages in the supply chain.

Table 9.2 The impact (+ pros and – cons) on the three pillars of sustainability, of the fully loaded secondary packages.

	Planet	People	Profit
Producer	+ Higher packaging fill rate	– More complex package handling	+ More products to the market
	– More material due to the insert		+ Better stability on the shelf
			– Investment in new equipment
Transport	+ Higher resource utilization	– Fewer truck drivers needed	+ Higher resource utilization
	+ Fewer transports		
Distribution Centre	+ Less warehouse space needed	+ More efficient storing	+ Less packaging material to handle
		– Fewer people needed	+ More products per delivery
Distribution	+ Higher resource utilization	+ More efficient handling	
Retail	+ Less packaging material to dispose	+ Less unstable packaging to deal with	+ Increased display efficiency
	+ Fewer orders	+ Better stability on the shelf	
Consumer		– More difficult to remove tubes	

9.2.4 Supply chain implications

When handling and distributing ACTA toothpaste, there are several potential improvements that can be gained by rethinking how the tubes are apportioned from 12 to 24 and packed together (Table 9.2). From an environmental perspective, the increase in fill rate per secondary packaging decreases the transport volumes, especially when moving the goods from Holland to Sweden, but also in distributing them to the different retail outlets. The transport of the product by the consumer from each pharmacy is not affected, since they usually do not buy more than one or a few primary packages at a time. The increased number of tubes per pallet also means that less space is required for storage in the Cederroth factory and the Cederroth warehouse. Since the pharmacies normally only order one secondary package at a time, the effect on their storage space will be insignificant.

Nonetheless, the suggested double stacking requires a slightly more complex secondary package and also a slightly more complex process when filling the secondary packages at the Cederroth factory. The more complex package might lead to slightly more handling time when replenishing products in the pharmacy. However, the quantities of packages to handle will be only half. The pharmacies will have to change their order size: today they only order 12 tubes (one current secondary package) at a time. With the double stacking solution, one secondary package will contain 24 tubes. This also leads to less handling times.

Case acknowledgements

Special thanks to Torgil Bråberg, Naja de la Cour, Daniel Johansson and Marcus Svensson for the initial analysis and the suggested improvements in this case.

9.3   The quantity of bottles in boxes

Case by Fredrik Nilsson and Daniel Hellström

This is an explorative case focusing on the apportionment ability of packaging and is based on two studies related to wine and liquor for one domestic and one international market. In the wine and liquor business there is a long tradition to distribute in packages of six and/or twelve bottles. If you ask people in the industry, "Why 6 or 12 bottles in a box?", they will most likely answer, "We have always had it that way". Thus, it is interesting to explore this tradition and standard.

9.3.1   The wine packaging system

South Africa has been the country with a long tradition in wine production and wine exports. The South African wine industry contributes 10% of GDP and therefore plays a major role in their economy and respective Agricultural sector (The Wine Institute, 2014). Two mid-sized South African wine cellars have been studied: the Simonsig winery and the Koelenhof. Both produce wine for the domestic market, as well as for export to Europe and the US predominately.

Most wines are sold in glass bottles and sealed with corks. However, an increasing number of wine producers are using alternative closures such as screwcaps and synthetic plastic "corks", as well as new packaging materials such as plastics.

The primary package for both wine producers is a glass bottle. The outer glass bottle diameter is approximately 76.3 mm compared to the inner being 63.1 mm. The total bottle height is 276.6 mm and weighs approximately 1.35 kg when filled. The glass bottle itself without contents weighs about 600 grams. The outer screw cap used is made of tin, while the inner part is a mixture of Poly-Vinyl Dichloride-PVDC and white polyethylene. In addition, a paper based product label is attached to the bottles.

A cardboard box measuring $324 \times 243 \times 298$ mm (L × W × H) is used as the secondary package due to its relatively low cost, impact resistance and handle ability. Its weight is approximately 200 g. When holding a maximum of twelve bottles, its weight increases to an average of 18 kg. Two polyurethane dividers separate and increase primary package stability.

The tertiary package consists of a 2-way reversible wooden pallet on which 56 or 60 secondary packages (14 or 15 boxes per layer) are stacked and covered with shrink wrap. The shrink wrap aids batch stability and protects the units from moisture absorption. The use of wooden pallets efficiently contributes to bulk product stock movement.

9.3.2 The wine supply chain in South Africa

Simonsiig and Kolenhof produce the wine and then sell it through domestic wholesalers or international exporters. After filling, the bottles are placed by hand into cardboard boxes (12 units per box) containing dividers (2 × polyurethane) and packed onto pallets by hand (56/60 boxes per pallet). The pallets that are used are 2-way pallets and not fully reversible. One pallet weighs 30 kg. Forklifts take these pallets of finished goods to the warehouse after the cube has been shrink wrapped. In the case of Simonsig, they have their own trucks to transport the cubes to the wholesaler, while Koelenhof's wholesaler picks up ready pallets of wine at their distribution centre. The wine is then distributed to outlets; in our study the retailers Par and Pick'n Pay on the South African market.

9.3.3 Improvement potentials identified

Interviews and packaging scorecards that were carried out in these two supply chains revealed that the tertiary packaging, the pallet, scored high, meaning that all actors were satisfied with its performance. The secondary packaging resulted in a different score, showing that the retailers in both chains were less satisfied with its performance. It was raised by the retailers that the weight of the secondary packages made handling difficult and the fact that on most shelves in the stores the wine was put in two rows of five per row. Hence, from every package they opened they had to put two bottles back and replenish them at a later stage. The fact that glass bottles are the primary packaging also affects the retailers in that they are heavy as well as brittle for retail personnel to handle.

Evidently, apportionment appeared as a potential area of improvement. For wine this related to the amount of bottles per cardboard box, and was something found to be an important performance aspect for the retailers but also for the wholesalers. The fact that the retail stores often need ten bottles of each when replenishing and the ergonomic factor of a box is that it weighs 2.7 kg less provides interesting insights into an area that is not often explored in the wine and liquor business, namely the traditions. The ergonomic perspective, both at the wholesale's distribution centre and at the retailer, is to increase the handle ability of the secondary package by inserting perforated handles on the side of the cardboard box to make carrying easier. This minor adjustment may have immense impact on time when it comes to handling the boxes at the various points in the supply chain.

9.3.4 The *Absolut Vodka* packaging system

The bottle of *Absolut Vodka* is a beautiful medicine look-a-like glass bottle from the 16th century with a cap made of aluminium. The bottle is recognized as the *Absolut Vodka*'s brand and plays a key part of the exclusiveness and market success of the product. The bottle is available in various sizes and colours, nonetheless they all have the same famous shape. The investigated bottle is one of the most common and contains 700 ml 40% vodka. This bottle measures 85 mm in diameter, is 225 mm high and weighs about 0.57 kg and 1.3 kg when filled. *Absolut Vodka* is one of the world's largest brands of liquor sold in more than 126 countries.

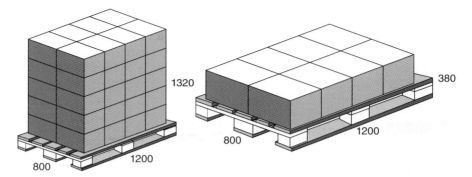

Figure 9.7 Illustration of the twelve pack box system configuration.

Typically, twelve bottles are packed in a secondary package resulting in a total weight of about 15.5 kg. The secondary package is a wraparound made out of C-fluted (3 mm) corrugated cardboard in several layers where sides and gable ends are glued. The secondary package measures 360 × 265 × 235 mm (L × W × H), consists of one 185 g outliner with an offset print, 140 g of half chemical fluting and one 175 g of white top liner on the inside. It also contains cardboard dividers to keep the bottles separated, making sure they do not get damaged while being transported or handled. For 700 ml 40% vodka, the annual production output is approximately 2 million bottles, which means that about 170,000 twelve-pack boxes are delivered from the factory every year.

The packaging system is made up by a EUR pallet with five layers, where each layer contains eight secondary packaging (Figure 9.7). A plastic wrap is used to protect the boxes from moisture and dust. It also gives some extra stability while being transported and stored. Plastic straps are then used around four of the layers for some extra security.

9.3.5 The *Absolut Vodka* supply chain

The Absolut factory in Åhus is highly automated. Receiving bottles, washing them, filling, labelling, packing, wrapping and loading onto trucks is almost totally automated. At distribution centres the secondary packages are picked out and put on mixed pallets for distribution to Systembolaget stores. Systembolaget has a monopoly in Sweden when it comes to selling liquor to private end consumers. At the stores, point-of-sale data is used in the ordering process and the *Absolut Vodka* product is delivered in twelve-bottle boxes twice a week. When received, the boxes are stored on the storage floor without any kind of pallet. The employees can then use a vacuum lift to handle a box. However, talking to employees indicates that the vacuum lift is not used as much as one could wish, since it takes a longer time. To replenish the store shelves, the twelve-bottle boxes are put on a trolley and each bottle placed by hand on the shelf.

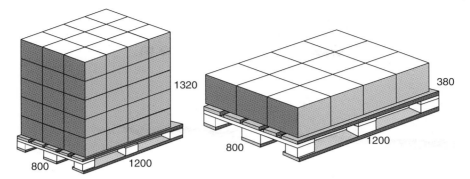

Figure 9.8 Illustration of a suggested nine pack box system configuration.

9.3.6 Potential packaging modifications and implications

Studying the interactions between the packaging system and the supply chain highlighted that there might be an opportunity to carry out improvements by considering the apportionment ability of packaging. Fewer bottles in the secondary package make it easier for the employees at the distribution centres and at the retail stores to handle the product. To enable this, the following packaging system configuration is suggested (Figure 9.8). With nine bottles instead of twelve, the weight decreases by 25%. Moreover, changing the apportionment enables an increase in fill rate of the pallet. The number of bottles per layer increases from 96 to 108. This means an decrease of transport and pallet handling by 12.5%.

When looking at the potential sustainability improvements, it is clear to see not only the direct positive results, but also some trade-offs. Having decreased the number of bottles in the secondary packaging from twelve to nine, results in more packaging material needed for the secondary packaging. If the currently used material is also used for the suggested box, there is roughly a 7% increase of material. Moreover, the factory needs to invest in its automated packing equipment to be able to pack nine instead of twelve bottles into a box. At the distribution centres a trade-off is that more units have to be picked which is associated with extra cost. However, when manual work is being done, each box has a reduced weight which is beneficial for the workers in the distribution centres. See Table 9.3 for a summary of the potential implications and the trade-offs from changing apportionment to nine bottles instead of twelve.

9.3.7 Concluding remarks: The quantity of bottles in boxes

This case shows that there are significant sustainability implications related to the apportionment ability of packaging. For *Absolut Vodka*, the suggested reduction from twelve to nine bottles per box significanly improves the ergonomic situation for workers, both in distribution centres and in retail stores (i.e. user-frendliness) as well as the fill rates. Regarding wine, similar advantages can be found when reducing the bottles from twelve to ten.

Table 9.3 The impact (+ pros and − cons), of the suggested nine pack box system for Absolut Vodka, on the three pillars of sustainability.

	Planet	People	Profit
Producer	− More packaging material		− Higher packaging material costs
	+ Less warehouse space needed		− Investments in equipment
			+ Less warehouse space needed
Transport	+ Less transports		+ Less transport costs
Distribution Centre	+ Less warehouse space needed	+ Easier picking	+ Less warehouse space needed
	+ Less forklift truck movement	+ Improved ergonomics	− More units to pick
Distribution	+ Less transports		
Retail	+ Less shelf space needed	+ Easier handling in storage	+ Less shelf space needed
	+ Less storage space needed	− More packaging material to recycle	+ Less products in storage

This case also shows that a perfectly fine and simple packaging system, containing a commonly used product (high volume), can be modified, based on its apportionment, to improve both fill rates and ergonomics and thus contribute to sustainable development. As mentioned earlier, the packaging system for these products are well-functioning and well-designed. However, the cases show that there are potential sustainability improvements to be made. Consequently, one learning outcome is that even if the traditional and standard packaging design works well and has done so for a long time, it is probably time for a change, since there are always improvements to be made. Do not let tradition set the standard and remember to question the most obvious "truths" in your supply chains and about your packaging system.

Case acknowledgements

Special thanks to Jennie Säwenvik, Melker Cullin and David Kollberg for their contributions to the interviews in the *Absolut Vodka* case.

References

Buchspies B., Tölle, S.J. and Jungbluth N. (2011), *Life Cycle Assessment of High-Sea Fish and Salmon Aquaculture*. Practical training report 25.05.11, ESU-services Ltd., fair consulting in sustainability, CH-8610 Ulster.

Nathan L., Pelletier N.L., Ayer N.W., Tyedmers, P.H., Kruse, S.A. et al. (2007), Impact categories for life cycle assessment research of seafood production systems: Review and prospectus. *The International Journal of Life Cycle Assessment*, 12(6), 414–421.

Nilsson K. and Lindberg U. (2011), *Klimatpåverkan i kylkedjan – från livsmedelsindustri till konsument" (Climate Effect of the Cold Food Chain – from the food industry to the consumer* – the authors translation*)*. National Food Agency, Sweden (Livsmedelsverket), Report 19-2011.

Svanen (2011): www.svanen.se

The Wine Institute (2014), *World Wine Production by Country 2014*: http://www. wineinstitute.org/files/World_Wine_Production_by_Country_2014_ cTradeDataAndAnalysis.pdf

10   User-friendliness

The package is the interface between the product and the users along the supply chain. This means that packages are handled by different users, for different purposes and in different ways. Therefore, user-friendliness is a direction that needs to be considered in packaging design.

Three cases were selected to represent the user-friendliness direction in the packaging design compass. They illustrate both improved use with packaging for people in their everyday work and for end consumers in daily life. Specifically, the cases illustrate redesigns of primary packages used by consumers, and primary and secondary packages used by employees at companies on a professional basis. The focus in these cases is mainly on the hurdles experienced by the users, but in one case also about regulations imposed by the authorities that have led to the redesigns.

The cases illustrate the effects these redesigns have had on sustainable development. The first case evaluates the positive and negative effects on consumers when the packaging of fever and pain reducing tablets was redesigned. The redesigned package is compared with the old one. The second user-friendliness case explores the potential effects of the invention of controlled delamination on improved usability. The case shows how in the future, controlled delamination can facilitate the opening of packages by hand that usually require tools. The third is a real-life case that describes a user-centred redesign of a business-to-business package. It is focused on the problems users had with the traditional package in their work situation and compares it with the redesigned package.

The cases are:

- Pharmaceutical packaging: Does size matter?
- Less frustration, less injury and less handling;
- TORK hand towels: Carrying to caring.

Managing Packaging Design for Sustainable Development: A Compass for Strategic Directions, First Edition. Daniel Hellström and Annika Olsson.
© 2017 John Wiley & Sons, Ltd. Published 2017 by John Wiley & Sons, Ltd.

10.1   Pharmaceutical packaging: Does size matter?

Case by Fredrik Nilsson and Annika Olsson

Medicine is targeted to different consumers with different needs; the need for the right medication, and the need to be able to handle the pharmaceutical packaging.

An often neglected area in the development and production of medicine is the packaging design and requirements; in particular, the requirements for protection and usability. Over the years, prescribed medication in the form of pills have been packed in paper bags, plastic jars, glass jars and recently, blister packs. Blister packs are commonly used in Sweden, but pills are also packed in plastic containers, which is more common internationally. Pharmaceuticals are regulated by national and international laws to ensure that users are not toxified or misuse the drugs. When it comes to the packaging regulations for pharmaceuticals, they are mainly about the graphics and information provided on the package.

Up until 2009, the market for pharmaceuticals was completely regulated by the state in Sweden, and medicine could only be purchased in state owned pharmacies.[1] In 2009, the market was deregulated and since then the most commonly used non-prescription or over-the-counter (OTC) pharmaceuticals are also sold in self-service grocery stores, gas stations and 24-7 after-hours stores. One of the largest volume products in Swedish pharmacies at the time of deregulation was paracetamol, a painkilling and fever reducing pharmaceutical under the brand name *Alvedon*. The attention was drawn to this product due to its high volumes and the shift in sales channels after the monopoly was lifted. It would, therefore, be a good illustrative case about packaging design and its effects on user-friendliness. It is further interesting to look into the supply chain implications of the redesigns.

10.1.1   The *Alvedon* supply chain

Before deregulation, the supply chain for *Alvedon* started at AstraZeneca's factories in Europe. In Sweden, Tamro was the primary wholesaler for *Alvedon* and along with Kronans Droghandel, another wholesaler, distributed *Alvedon* to the different pharmacies on the retail market. After deregulation, Kronans Droghandel expanded their supply chain to include retail stores. This resulted in the existing retailers expanding their supply chains upstream, as did Apoteket Hjärtat who now have their own supplier, Apo-Pharma. So the two initial wholesalers now have more competitors. At the same time, the state-owned Apoteket has developed its own brand of paracetamol, *Apofri*, in a new package. This means that there are similar products on the market for consumers to choose from that go by different names.

---

1   Apoteket AB (2009), *Organisation*, http://www.apoteket.se

10.1.2 The *Alvedon* packaging system

Before deregulation, the packaging system for *Alvedon* consisted of plastic blister packs that contained 10 pills of 500 mg each. Three such blister packs (30 pills in all) were enclosed in a carton box that measured $90 \times 55 \times 25$ mm, making up the primary packaging. The blister packs were made of transparent plastics sealed with an aluminium bottom layer. In addition to the brand name, the information on the primary package included the paracetamol content and quantity, batch number, expiry date and the recommended dosage according to age and weight. Inside, there was an informative leaflet with the required medical data, such as the side effects of paracetamol, and further considerations on recommended dosages. The secondary packaging consisted of a larger carton box that contained 42 primary packages, placed in 3 rows of 14 primary packages of *Alvedon*. The secondary packages were then placed on a pallet as a tertiary packaging.

After deregulation in 2009, each primary package held only two blister packs with 10 pills each, which meant 20 pills per consumer package. In 2012 the new producer, GlaxoSmithKline (GSK), decided to replace the bottom layer of aluminium with plastic to have a single material inner package, which would improve the environmental profile of the primary packaging. This meant that the entire blister pack is now produced solely in plastics, on both the top and bottom layers. The new measurements of the primary package are now $85 \times 70 \times 17$ mm. The secondary packaging box contains 24 primary packages with 20 pills per consumer package.

At the distributor, a process of repacking takes place. The tertiary packaging is removed and the secondary packages are placed in a blue plastic container along with the other pharmaceutical products that make up the order to be delivered to each pharmacy or retail outlet. The design of the secondary packaging before deregulation meant that the minimum distribution of this product was 42 boxes of 30 pills each. After deregulation, it was 24 boxes of 20 pills each.

10.1.3 Implications from a user-friendliness perspective

This case focuses on improvements in the *Alvedon* blister pack based on packaging scorecard analysis (Olsmats and Dominic, 2003) where user-friendliness received the lowest score, and also on the effects from changing packaging size after the deregulation. A look at the consumers' results in the packaging scorecard showed that many senior citizens do not like pharmaceuticals in blister packs because it is difficult for them to get the pill out of the blister. This specific problem was also ranked eighth from the top out of more than a hundred distinct issues concerning consumer complaints about pharmaceutical packaging in a study conducted in 2002 for Apoteket.[2]

On the other hand, both consumer safety and product protection were weighted high on the packaging scorecard by all the consumers who participated.

---

2 The Swedish Rheumatism Association 2002

They felt the packaging should be childproof and yet accessible for elderly citizens in particular, and other adults in general. The actors in the supply chain who scored the packaging system for *Alvedon*, asserted the needs for product protection, optimal stackability, volume and weight efficiency. The producer of *Alvedon* also pointed out the financial dimension where potential improvements must be easy to implement, cost effective and not cause any damage to the brand but still distinguish it from other brands.

From a user perspective the dominating issue is the difficulty in getting the pill out of the blister, especially for people with a reduction in muscle strength and hand function. One improvement that could be easily achieved is to offer Alvedon as a 1,000 mg pill, since this is the standard dose and the one most commonly used by an adult (instead of the current solution of taking two pills at the same time). For adults, 1,000 mg pills would occupy less space in the blister packs compared to the two 500 mg pills, since the 1,000 mg pill does not need to be bigger; instead, it can have a higher concentration of paracetamol. The 1,000 mg pill solution also would result in greater volume efficiency in the outer primary package because the pills in the blister packs would be closer together. This change would retain the primary package's stackability and it would have the same measurements for the secondary solutions as before. This solution is feasible from a user-friendliness point of view and also because more paracetamol (double dose) can be distributed in packages of the same size as before.

However, the solution can be questioned in terms of the risk for toxicity with 1,000 mg pills instead of 500 mg pills. This risk was identified after the monopoly was lifted in 2009. This deregulation certainly brought improvements for consumers because they were able to purchase products in more locations and at stores that had longer opening hours. But the increased availability might have stimulated over usage.

As long as *Alvedon* was only sold through state owned pharmacies, the primary package contained 30 pills in blister packs in consumer carton packages (Figure 10.1).

After its release in ordinary supermarkets, the Swedish Poison Information Centre alerted the Swedish Medical Products Agency to the increase in abuse of paracetamol, especially among young people. This resulted in a new packaging regulation from the Swedish Medical Products Agency to reduce the consumer packaging size so that it contained no more than 20 pills. The package was redesigned with the same kind of protection but it meant "more packaging material per packed product" (Figure 10.2). From the consumer perspective, the price of one package was kept the same, which meant that the price per pill increased by approximately 50%, resulting in a solution that became more expensive both in terms of packaging material usage and the price to the consumer.

Whether this design change has reduced the number of overdoses is unknown, since there are no restrictions on how many primary packages consumers are able to buy, or on how often they are allowed to buy them. What is clear, however, is the increase of packaging material per pill, and less efficient space utilization

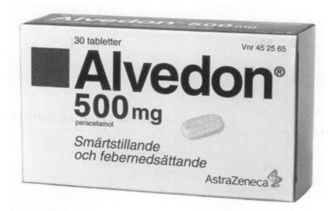

Figure 10.1 *Alvedon's* primary package before deregulation.

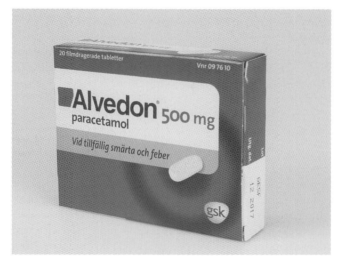

Figure 10.2 *Alvedon's* new primary package.

resulting in fewer pills per pallet in transport and more expensive solutions resulting in higher prices for the consumer.

10.1.4 Sustainability implications

The proposed solution of making a more concentrated pill (1,000 mg) to better fit the dosage recommendations for grown-ups was never implemented. It would have been a more efficient packaging solution because it reduced the volume in the packages by half. It would have involved less transport, less handling, less order quantities but possibly more differentiation in terms of packages with different pill concentrations for different target groups. This would have resulted

Table 10.1 The impact (+ pros and – cons) on the three pillars of sustainability, for *Alvedon* pill packaging with the current 20 pills per primary package instead of the previous 30.

	Planet	People	Profit
Producer	– More packaging material per packed pill	– More handling	+ Higher margin per package sold
Transport	– More transport volume	– More handling	– More volume to transport
			– More space required in transport
Distribution Centre	– Less volume efficient	– More handling	
Distribution	– More transport volume	– More handling	– More volume to transport
			– More space required in transport
Retail	– More secondary packaging to discard	– Increased replenishment	+ Higher margin per package sold
			– Increased cost for handling and replenishment
Consumer	– More packaging material to discard	– More handling	– Increased cost
		– Fewer pills per package	
		+ Less risk of toxicity	

in more complex distribution solutions and the risk of mixing different concentration if the packages were very similar. The most important aspect of this proposed solution was the convenience for the user in handling half the number of pills for the same pain reduction effects, thus eliminating the usability hurdle reported in the scorecard survey. It would have been better, however, to innovate a package solution that made it easier to access the pills, even with limited manual dexterity, independent of the pill's concentration. The drawback of the proposed solution is clearly the risk for toxicity and overdosing that ultimately led to the redesign of the current package of only 20 pills per consumer package in the first place. This new solution, on the other hand, has had the opposite effect from a sustainability point of view: more transports due to less efficient packages; more handling per package due to fewer products in each package; more packaging material per pill; and yet the same solution for consumers when it comes to accessing the pills from the blister. So from a usability point of view the new solution has made no change for the consumer and only more handling per pill for all actors who deal with the packages in the supply chain (Table 10.1).

Case acknowledgements

Special thanks to Mazen Saghir who contributed with input on regulations and redesigns on the packages after the introduction of *Alvedon* in ordinary retail stores.

10.2 Less frustration, less injury and less handling

Case by Annika Olsson

Consumers continuously complain about packages that are hard to use, especially when you want to open them. One category that is most frequently mentioned in these situations is the plastic blister packages for products like electronic devices.

The trade-off that the packaging industry struggles with in consumer products that are valuable and "attractive to steal" is to find packaging solutions that are easy to open when legally purchased but are impossible to steal from, or to open in order to replace the contents with counterfeit products. Similar problems are encountered in the pharmaceutical industry.

10.2.1 The controlled delamination invention

One potential solution to this trade-off is the invention of controlled delamination materials (CDM). CDM for packages was the brainchild of an entrepreneur employed at Swedish Stora Enso. It has since been further developed by a couple of inventors including the initial entrepreneur, into a spinoff company, Exonera.[3] The idea is based on electrically controlled adhesives, originally developed for attaching test equipment on airplanes. The idea behind the application of controlled delamination materials for packaging is to connect the material to a circuit, either through conductive packaging material or through conductive lines printed on the package. Conductive packaging can be produced by using standard processes for electronic printing. The conductive packaging is then joined to a glue spot that seals the item or that seals different conductive packages together. Once the circuit is closed, delamination takes place, the adhesive effect is lost and the seal breaks (*Packaging Professional Magazine*, 2006). The circuit can be closed, for example, by pressing a switch made from printed conductors that can be activated on contact with human skin.

The applications of this invention in the packaging industry are two-fold:

1) ***Application 1***: The use of CDM with built-in internal sources for easy opening of tightly sealed primary packages such as blister packs, or hard-to-open laundry detergents boxes or soft drink cans.
2) ***Application 2***: The use of CDM to hold primary packages together and replace secondary packaging in items like flexible cartons or cans that are frequently used for carbonated and non-carbonated soft drinks.

The inventors focused on bringing Application 1 to market first because of the project logic and time to market. They planned to achieve this by the use of a pre-made laminate that was already commercially available.

For expensive and attractive consumer products, such as consumer electronics in blister packs, this first application of CDM provides theft prevention and authenticity of product as long as the packages are sealed in the retail outlet. The packages can be easily "unlocked" by applying a little electric power at the

---

3 www.exonera.com

Figure 10.3 Consumer packages suitable for the CDM application.

checkout counter after paying for the item. When the package passes the electric power source, the controlled adhesive releases and the package opens up.

For commodity products such as detergents, the packages need to have tight seals that do not break before they are opened in the home of the consumer. The CDM solution can be applied to easily open the package using the conductivity from human skin as a switch (Figure 10.3). In a similar way, the CDM can enable the easy opening of pull rings on soft drink cans, using the humidity of human skin as a switch to initiate the release process. In these examples, switches with printed conductors can be used and activated by skin conductivity. Pressing one's finger on the switch releases the CDM's seal and opens the lid of the package. This application will facilitate the opening of packages from a consumer point of view.

The primary potential of Application 2 is in its aim to tightly connect primary packages together (Figure 10.4). The CDM solution would then consists of external battery power sources for secondary distribution packages. Connecting the primary packages with polymer glue makes it possible to transport a set of such primary packages with a protective top cover and cardboard bottom as secondary packaging to the retail stores. Releasing the CDM at multiple points will then remove the top cover and separate the items individually or in batches for shelving. Application 2 will reduce the amount of secondary packaging material, which usually consists of a wrap-around cardboard box. It will also reduce distribution costs by significantly shortening the handling time when transferring products from distribution packaging to store shelves. At present, unpacking the distribution packages and placing the consumer packages on retail shelves is time-consuming and a hurdle for the store personnel in packaging material handling. In addition, there is a considerable amount of waste material that has to be removed from the store and turned in for recycling by the employees. Using CDM can reduce in-store handling times and improve store logistics.

10.2.2 CDM sustainability implications

The main advantage of using controlled delaminated materials when it comes to primary packages is on the side of people. But CDM technology also has positive

Figure 10.4 CDM applications for replacing traditional secondary packaging.

effects on the side of profit. The existing plastic blister packs for consumer electronics and other products cause much frustration for consumers when they go to open their purchased products: firstly, because they need to use tools, and secondly because of injuries. Some consumers hand them over to the cashier, insisting that he or she open the package before they leave the store. This potential CDM solution can very well decrease consumer frustration and increase customer satisfaction, which leads to a better position for the producer and better profits down the road.

When it comes to reducing theft and counterfeit problems in the supply chain, CDM technology has potential advantages for sustainability, both environmentally and economically. By reducing the number of missing, stolen or misplaced products, less product production is needed, fewer claims need to be handled, and fewer products returned. This will potentially lead to fewer transports and in that way have a positive environmental effect. This all further means positive economic effects predominantly on the product side but also on the packaging side.

The main advantage of the CDM solution in replacing secondary packaging is from the planet perspective. This is because it leads to less packaging material use in several steps of the value chain. And by eliminating the secondary packaging, the products are visible in distribution centres, reducing the number of handling mistakes. However, it is not necessarily the case that the limitation or elimination of secondary material will be to the benefit of the primary package, because this means that it will be less protected in transport and handling. In order to introduce a CDM system, one needs to evaluate the potential negative effects of an increase in destroyed primary packages, which would have a negative impact from the planet perspective (Table 10.2).

It is worth mentioning that the CDM technology is still in the developmental phase and not yet on the market. But it can be a potential solution for more

Table 10.2 The impact (+ pros and − cons) on the three pillars of sustainability, of primary packaging with CDM applications.

	Planet	People	Profit
Producer	+ Fewer lost products	+ Fewer lost or replaced products	+ Innovative customer solutions
	+/− Effect on production cost	+ Greater customer satisfaction leading to more sales	− Investment in new packaging solutions and equipment
Transport	+ Less transport		
	+ Fewer return transports		
Distribution Centre	− May result in more destroyed primary packages due to less protection	+ Easier to see content if secondary package is replaced	+ More efficient handling
Retail	+ Lower product replenishment time	+ Authenticity assurance	+ Less theft
	+ Fewer wasted products	+ Fewer claims to handle	
	+ Less secondary packaging material to handle		+ Less fraud
Consumer	+ Less injuries	+ Authenticity assurance	
	+ Less frustration at handling	+ Less frustration with openability	

sustainable packaging in the area of optimized use and reduced handling times in the entire supply chain, including the homes of the consumers.

Case acknowledgements

Special thanks go to Lars Sandberg, the entrepreneur behind this invention, for sharing his knowledge and insights on the case, and for reading through and giving feedback on the case writing.

10.3   TORK hand towels: Carrying to caring

Case by Daniel Hellström

Every day, about 4 million cleaners go to work in Europe. They keep our workplaces, arenas, schools, convention centres, shopping complexes, hotels and hospitals clean. The work of cleaners is tough and the rate of sick leave and work-related health problems is higher in the cleaning industry than in many other sectors (Scheil-Adlung and Sandner, 2010). According to the European Agency for Safety and Health at Work (EU-OSHA, 2009), one major health problem among cleaners is musculoskeletal disorders. These include injuries and pain in areas

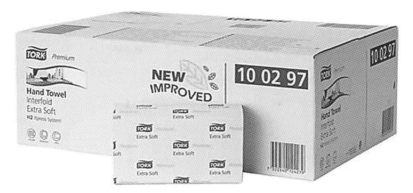

Figure 10.5 The old TORK hand towels primary and secondary packaging. The carry handles are not visible.

such as the back, neck, knees or wrists. The risk of musculoskeletal disorders increases with heavy lifting, carrying and twisting, all of which are common for cleaners (Kumar and Kumar, 2008).

An important part of cleaning work is the replenishment of hygiene products such toilet paper, paper hand towels and soap. These products arrive in packages, and the handling of these packages is a common everyday work activity for cleaners. For paper hand towels alone, approximately 2 million packages are handled every week by cleaning workers in Europe.

TORK is a brand of SCA (Svenska Cellulosa Aktiebolaget) for away-from-home hygiene tissue solutions. TORK products such as toilet paper, paper hand towels and soap are found in millions of washrooms around the world. In order to make the work of cleaners easier and more convenient, SCA, a leading global hygiene and forest products company, initiated a study (Grennfelt, 2012) to improve the packaging solution for their TORK brand of hand towels by uncovering tacit needs and wishes from the users involved.

10.3.1 The previous TORK packaging system

Many of the away-from-home hygiene products on the European market are distributed in corrugated board boxes. The previous secondary packaging box for the TORK hand towels (Figure 10.5) was a regular slotted cardboard box with carry handles on the short sides, containing 21 bundles (7 × 3) of towels. The box dimensions were 215 × 395 × 600 mm resulting in four boxes on each pallet layer. Each box was sealed with transparent tape.

10.3.2 The TORK supply chain

This supply chain is entirely a business-to-business transaction. Nonetheless, the hygiene products eventually end up being used by the general public and are handled in the last part of the chain by the cleaners. The supply chain consists of a producer, wholesalers and facility service companies or any other company or organization (i.e. facility). These actors have various warehouses and storages. The product flow between the actors varies, depending on the supply chain set-up. For example, hygiene products are mainly transported from the producer

Figure 10.6 Generic supply chain for TORK hand towels.

to wholesalers and then to the end customers — the facilities of companies and organizations. Sometimes the transport is from the wholesaler via a facility service company or directly from the producer to the facilities. The overall supply chain is illustrated in Figure 10.6.

From a packaging point of view, all cardboard boxes are transported to the warehouse(s) on pallets, where they are picked and sent to the facilities in mixed product shipments or as whole pallets. The facilities have different levels of storage. It is common to have a first storage room and several second storages. The cardboard boxes are initially transported from the facility's goods reception to the first storage room. Then the boxes are manually lifted and carried out of first storage and transported to the second storage rooms – in larger facilities involving longer distances. Second storage rooms are often very limited in space. There, the box is opened, unpacked and disposed of or used to transport the hand towels to the place of usage. The cleaning workers have to use a tool to open the taped cardboard boxes, often a knife.

Subsequently, cleaners spend a considerable amount of time and effort in handling the boxes. Identifying, lifting, carrying, opening and disposing of boxes are a big part of their daily work. This handling is tiresome and takes time away from the actual job of cleaning. The shape and size of the box is not made for lifting and carrying. Even though the actual weight of the box is light, you need to use both hands to lift and carry it due to its inconvenient size. This makes it very cumbersome to carry and causes difficulties when you need to open doors and walk up or down stairs. You can basically only carry one box at the time. When the box is finally empty, it needs to be unfolded and carried to the recycling area. The boxes are hard to unfold, and even harder to carry when unfolded.

10.3.3 Packaging evaluation and redesign

The SCA study resulted in various insights into the supply chain actors' daily jobs and interactions with the packaging system. The insights from the users — the cleaners — were that their daily work was tough and they did not want to lift and carry more than necessary. So if the packaging solution were easier to store, lift, carry and dispose of, their life at work would be much easier. Insights from the facility service companies were that they wished to reduce the number of days employees took sick leave. One action they could take to accomplish this was to reduce the physical burden of the employees. The facility service companies also wanted their staff to work in a more efficient manner in the

Figure 10.7 The improved TORK hand towels secondary packaging solution.

disposal of the package material. The insight from the producer was simply that if they changed the packaging solution to a better one, the service level and the acceptance at the end customer would increase, resulting in an improved competitive advantage. Based on these insights, the following six criteria for improvement were identified:

1) easy to handle, ergonomically improved;
2) easy to open;
3) easy to store and less space needed to store;
4) easy to dispose of;
5) improved visual appearance;
6) improved transport and stability; and
7) environmentally sustainable solution.

SCA Hygiene Products developed a secondary packaging solution based on a plastic bag with a handle on top that was assessed to be the best (Figure 10.7) and was launched on the market in 2009. The plastic bag is made of polythene and is recyclable. From a sustainability point of view, the environmental impact of the new plastic bag is equal or less due to lower material weight, even though it is replacing a material based on renewable cardboard. The content, physical dimensions and packaging cost are approximately the same as the previous solution.

10.3.4 Supply chain impact

The decision to change the secondary packaging solution had several impacts on the producer. An investment in new and more expensive packaging machines had to be made, but the storage space needed, and the receiving and handling of incoming packaging material decreased. Still, the producer was concerned about potential product and packaging damage throughout the supply chain. The new solution proved to be durable, even without the corrugated board.

For the cleaning workers, the plastic bag solution has improved ergonomics and efficiency. Many cleaners have expressed their satisfaction with the new packaging solution. The bag is much easier to lift and carry with the new handle. Previously, both hands had to be used when carrying, but with the single handle, cleaners have one hand free for opening doors or carrying a second bag. The plastic bag also has a perforation making it easier to open, without using a knife

Table 10.3 The impact (+ pros and – cons) on the three pillars of sustainability, of the redesigned packaging solution for TORK hand towels.

	Planet	People	Profit
Producer	+ Less storage space needed		+ Less storage space needed
			+ Less handling of incoming pallets
	+ Less handling of incoming pallets		+ Fewer pallets in the system
			+ Improved product display
	– Less use of renewable resources for packaging		+ Increased sales
			– New packaging machine
Transport	+ Less packaging damage		+ Less packaging damage
Wholesaler/ Distribution Centre	+ Less packaging damage		+ Less packaging damage
Distribution	+ Less packaging damage	+ Easier replenishment	+ Less packaging damage
Facility	+ Less storage space needed	+ Easier lifting	+ Less storage space needed
		+ Easier carrying	
		+ Easier opening	+ Improved work environment
		+ Easier unpacking	
		+ Easier disposing/ recycling	+ Improved work efficiency
		+ Easier replenishment	

or other tool. The plastic bag requires less storage space when the towels are taken out of the package. This results in more available storage space for other things. According to the cleaning workers, one of the most appreciated impacts of the plastic bag solution is the easier disposal of empty packaging. The cleaners no longer need to spend time and energy unfolding and carrying the boxes to the recycling area. One additional benefit of the plastic bag is its visual appearance and transparency, which makes the product easier to pick out and identify.

10.3.5 Concluding remarks: Carrying to caring

This case shows that user-friendliness goes beyond primary packaging and end consumers, by specifically addressing secondary packaging levels in business-to-business transactions. It also shows that there are significant supply chain sustainability implications related to the user-friendliness of packaging. For TORK hand towels, the new secondary packaging solution has resulted in improved ergonomics (people) and efficiency (profit), influencing the sustainability across the entire supply chain (Table 10.3). From a profit perspective, the producer had to invest in new and more expensive packaging machines but gained benefits such as handling less incoming pallets with packaging material. The facility organizations and service companies gained the benefits of enabling their cleaners to work in a more efficient manner and of reducing the physical burden on the employees. This is connected to the people perspective, which is the major benefit of the new improved packaging solution. It improves life at work for Europe's 4 million cleaners by making it more ergonomic and easier to identify, lift, carry, open and dispose of the approximately 2 million packages handled per week of paper hand towels alone.

Case acknowledgements

Special thanks to the packaging innovation and design team at SCA Hygiene Products, especially Kristian Grennfelt (Global Brand Innovation Manager) and Hans Wallenius (Global Innovation Director), for sharing their reflections on this project. Thanks to Bengt Järrehult as well for his insightful comments.

References

Apoteket A.B. (2009), *Organisation*: http://www.apoteket.se

EU-OSHA (2009), *The Occupational Safety and Health of Cleaning Workers*. European Agency for Safety and Health at Work (EU-OSHA): https://osha.europa.eu/en/publications/literature_reviews/cleaning_workers_and_OSH

Grennfelt K. (2012), *Creating a Better Working Environment for Cleaning Workers throughout Europe*: http://www.sca.com/PageFiles/57111/SCA-TORK-Report-2012.pdf

Kumar R. and Kumar S. (2008), Musculoskeletal risk factors in cleaning occupation – a literature review. *International Journal of Industrial Ergonomics*, 38(2), 158–170.

Olsmats C. and Dominic C. (2003), Packaging scorecard – a packaging performance evaluation method. *Packaging Technology and Science*, 16, 9–14.

Packaging Professional Magazine (2006): (http://www.iom3.org/news/controlled-delamination-materials-technology-separate-packs-electronically)

Scheil-Adlung X. and Sandner L. (2010), *The Case for Paid Sick Leave*. World Health Organization: http://www.who.int/healthsystems/topics/financing/healthreport/SickleaveNo9FINAL.pdf

11   Information and communication

The fast development of information and communication technologies and global digitalization of information, assigns new needs and requirements on the packaging system. Three cases are presented to illustrate that the information and communication related to packaging systems is one of the six fundamental directions in our compass for developing and managing sustainable design. The cases focus on different packaging system components and their impact on sustainability across supply chains.

The first case explores an intelligent time and temperature indicator for packages in the fresh and chilled food distribution segment. The case illustrates an invention, not yet released on the market, that can help reduce food waste and make food distribution more reliable and safe. The second case compares and emphasizes the different information and communication needs on primary packaging for mobile phones based on two market viewpoints. The third real-life case demonstrates that packaging can be used as an instrument for gathering accurate, reliable and timely information that enables traceability of returnable packaging. This results in a reduced need for investment in expensive returnable packaging systems and more sustainable supply chains.

All three cases show that there are significant sustainability implications related to the information and communication aspects of packaging systems and that it is vital to consider these in the management of sustainable design.

The cases are:

- How do you know if the milk is sour? An innovative sensor technique;
- Mobile communication through the package;
- Roll containers for diary cartons: Connecting atoms and bits.

Managing Packaging Design for Sustainable Development: A Compass for Strategic Directions,
First Edition. Daniel Hellström and Annika Olsson.
© 2017 John Wiley & Sons, Ltd. Published 2017 by John Wiley & Sons, Ltd.

To complement the three cases, an additional section further discusses packaging as the silent salesman. This involves the sales and marketing function of communication. Here, aspects such as logotypes, user instructions, shelf display, and labelling and their impacts for sustainable development are discussed. This additional section, 11.4, is entitled:

- What does the silent salesman do for the sustainable society?

11.1 How do you know if the milk is sour? An innovative sensor technique

Case by Annika Olsson and Fredrik Nilsson

Today consumers heavily rely on the stamped "best before date" for evaluating whether their fresh food is still edible or if they should toss it in the waste. According to the European Commission, 90 million tons of foods are wasted annually. In the UK alone, almost one-third of the food purchased goes to waste, of which two-thirds still has good quality (European Commission; WRAP, 2007). The labelling of the best before date and use by date on food is one of the major reasons behind food waste (European Commission; Loxbo, 2011; Stockholme Consumer Cooperative Society, 2009). In Sweden, approximately 33% of people throw their food away because the best before date has expired (Stockholme Consumer Cooperative Society, 2011). But is this the best way to find out? And is it the most sustainable way to treat fresh products?

All food products deteriorate over time and have a given shelf life that indicates the level of quality. Different food categories have different sensitivities to deterioration processes and in general, fresh and chilled categories are more sensitive and thus more demanding when it comes to transportation, distribution, storage and planning. The two most important factors concerning the shelf life of fresh and chilled food products are time and temperature. Deteriorating changes that take place in food are most frequently temperature dependent and occur at a slower rate at lower temperatures (Hernandez, 2001).

All food products in Sweden have to be labelled with either a best before date or a use by date. The latter is used for more sensitive food like fresh meat, fish and chicken (Livsmedelsverket, 2004a). The best before date is set so that the food, if stored under the right conditions and in an unbroken package, will be consumable some additional time after the given date (Livsmedelsverket, 2004b). Since food producers have to add a margin of safety to the best before date, this results in an unnecessary waste of food, both by the consumers in their home and at retail locations, since the best before date is meant to be used as a guiding principle for when to take products off the shelves and when to trash them.

One way to handle the food waste and the problems that arise because of the inaccuracy of the best before and use by dates is packaging solutions that involve time-temperature indicators. These can measure, communicate and inform supply chain actors and consumers about the "real" shelf life of actual products rather than predictions that have safety margins that create planning and waste problems in the whole food chain. The time-temperature indicator presented in this case was invented to help consumers and other supply chain actors to identify the real shelf life of a product.

One of the earlier Swedish initiatives – a time-temperature indicator – is explored in this case, to show the potential benefits of packaging technology that can communicate and inform the supply chain actors, including consumers, about the actual shelf life instead of the predicted shelf life dates stamped on the packages. The inventor's idea was to come up with an active tag for food quality measurements that could be used for chilled distributed food. The first version was for producers, distributors and retailers, but in later versions, it was also intended to target consumers.

The tag was made up of an electrical circuit combined with an enzymatic liquid solution. The tag could be read by a handheld scanner. The results of the measurements in the different parts of the chain were then logged into an Internet portal where all actors could follow the product and its distribution conditions on its way from the producer to the retailer. The measurements show the accumulated time and temperature exposure for a product from production to consumption. This is because shelf life quality is dependent on the total time a product has been exposed to a temperature that is too high. This biosensor tag is activated when applied to a package in the production filling line at the producer. After application the active tag accumulates the temperature and time exposure data of the product in temperatures above the one stipulated for the chill chain. This results in a more rigorous measure, taking accumulated time and temperature into account in comparison to using on-off sensors that turn red or indicate in another way when the product has been in the wrong temperature zone on one occasion.

The accumulated time-temperature relationship used in this invention represents the total area under the blue curve. When that area reaches a limit, indicated by the red line in the graph shown in Figure 11.1, the product has moved beyond its shelf life. Most other temperature indicators with on-off regulation

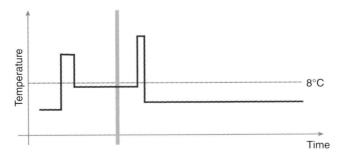

Figure 11.1 Graph of accumulated time and temperature peaks.

Figure 11.2 Temperature biosensor circuit and tag.

would sound the alarm when the first peak above the stipulated temperature was measured, meaning that the first time the product is out of its temperature zone the indicator turns red, and the product is reported as having gone beyond its shelf life. As for the graph in Figure 11.1, the first peak above the stipulated temperature of 8°C happened long before the product in reality had moved beyond its shelf life quality.

The sticker was further integrated with an RFID circuit and a bar code, which helped companies to trace and control the product through the supply chain. The tag had the biosensor hidden under the RFID and bar code sticker, as shown in Figure 11.2.

11.1.1 Implications of supply chain implementation

The benefit of introducing these bio-sensing tags for accumulated temperature exposure is that it can help supply chain actors measure the temperature exposure of the product from production to retail. This assists the actors in identifying products that need to be taken off the shelf or sold earlier due to an expiring shelf life. It can also help to identify whether products that have passed the best before date are still edible.

Prior to starting development, initial discussions with producers and retailers proved the innovative biosensor to have business potential. These actors could see the added value in having this type of intelligent sticker on a package. The producers appreciated the added value of being able to secure the temperature at storage right after production. They also saw an opportunity to trace the supply of products throughout the food supply chain by getting information about all temperatures from production to retail. The RFID tag with the bar code was also seen as value adding due to its properties to store all relevant data for the product in addition to the temperature data from the active circuit. This would make it possible to direct flows in a more efficient way, thus avoiding waste and increasing sales (Olsson, 2010).

The distributors saw the value of detecting products that have been exposed to a higher temperature than stipulated. This would enable them to redirect the orders and start distributing the products with the shortest shelf life first, rather than continuing to distribute according to the stamped best before date, resulting in less waste.

On the retail level the added value was the knowledge about the temperature exposure of products. This meant that they would know when to take products off the shelf when the exposure was too high. The added service could also be used to put products on the shelf in another order rather than relying solely on the best before date stamp to reduce waste of short-life products.

Even though all actors could see its potential value, the biosensor was never introduced, because no single actor was interested in taking on the investment costs that were most likely higher than those for on-off indicators with less complex technology behind them.

The market of time and temperature indicators (TTI) has increased since the closedown of the case company. Several companies provide TTI solutions to track chill chains worldwide, such as *TempTale* by SensiTech, *Smart Trace* by Smart Trace Online Monitoring and *MonitorMark*™ by *3M*™. However, most of these are used by food producers and distributors that perform random or addressed tests to monitor the quality of their chill chain distribution. They are not focused on food waste or, for that matter, better information to the actors in the chain to improve supply chain operations. One sensor that has received attention on the Swedish market is *Tempix*[1] – an indicator that according to its website is yet another on-off version. The *Tempix* website states:

> The Tempix temperature indicator will reveal whether a particular product has been handled at too high temperatures during its journey from the producer to the final customer. When you see the black bar in the Tempix indicator, you can be assured that the product has been kept at the correct temperature throughout the cold chain. Should the product have been exposed to temperatures above the recommended limit at any stage of its handling, the black bar will disappear from the window. The barcode will also be blocked out, meaning it cannot be read at the till. The indicator therefore acts as a guarantee that the product has been handled correctly from producer all the way to the store's cashier.[2]

The problem with the success of on-off indicators is that they most likely lead to an unnecessary increase of product waste, since these indicators guide consumers not to trust the product if it has been exposed to a higher temperature on one single occasion. But since the shelf life depends on the total time of temperature exposure, a single occasion may not be enough for the quality to be below the appropriate shelf life level. The intention of these types of indicators is good in the sense that they promote transparency all the way to the consumer. But the good intention can result in unnecessary waste of products, packages, indicators, transports and man hours, which in turn leads to higher product costs for the consumer. The *Tempix* indicator will even block out the black bar on packages that have been exposed. This disables the cashier's ability to scan them, and thus makes them impossible to sell.

11.1.2   Sustainability implications for the indicator invention

From a sustainable perspective, the different supply chain actors in the case described saw potential for reducing food waste by means of better control that was achieved by using the biosensor for accumulated temperature registration.

---

1 http://tempix.com/the-indicator
2 Ibid.

Producers could identify cooling issues in the production and storage states, distributors saw a potential to redirect the flow of items based on their shelf life, and retail saw the potential to relocate products according to their "real" shelf life. All these actions would reduce the waste of packed products and thus increase sales. With less waste, the different actors would also reduce the time required to handle items that needed to be taken out of stock or removed from the retail shelf.

Clearly, the introduction of biosensors incurs an extra direct cost per package. In isolation, this might not be a major problem and could be justified in relation to the added value it offers the actors. But it also results in indirect costs for machinery, handling time, and time to read the values at the different locations in the supply chain. The food producers may be the ones expected to carry the cost of investing in packaging and related items, but they are not expected to charge a higher price for the product (Olsson, 2010).

The sensor as such evidently increased the cost of the package system. That was the reason that its implementation did not come into being, even if the total increased value in the supply chain may well have increased with more sold

Table 11.1 The impact (+ pros and – cons) on the three pillars of sustainability, for packages with biosensor for temperature indication.

	Planet	People	Profit
Producer	+ Less use of packaging material due to less waste	+ Better control of the supply chain on the Internet platform	+ Better control and planning + Better potential in agreements with customers
	+ Less use of products due to less waste	+ Better transparency of supply chain More process time due to additional operations	– Investment cost in machinery – Extra cost per package
Transport	+ Better planning and control based on real shelf life		
Distribution Centre	+ Redirecting flow of items according to shelf life	+ Reduce the time to handle items that should be taken out of stock	+ Fewer returns
Retail	+ Better planning of display order of products	+ Better transparency and control of the supply chain + Reduced handling time of items that have passed expiration date	+ Less waste and thus better profit + Fewer returns
Consumer	+ Less product waste	+ If exposed to consumers they can make appropriate actions and reducing waste – Better safety and trust of brand	

products, fewer returns, fewer transports, less packaging material and reduced usage of raw material. It would definitely have been beneficial from the people and the planet perspective of sustainability.

The problems identified in the implementation phase indicate the problem of sharing risks, costs and also potential value among supply chain actors. It further indicates the problem of taking overall responsibility for a chain problem. The transfer of power from the several producers in the Swedish food industry to the few wholesalers and retail chains has further intensified the difficulties of sharing problems as well as business opportunities along the chain. This confirms that the potential implementation of the biosensor exposed the insufficient overall chain responsibility in the important area of securing safe food and reducing waste of non-expired food (Table 11.1) (Olsson, 2010).

Case acknowledgements

Special thanks go to Bengt Sahlberg, the former CEO of Bioett AB. Access to the development process, sensor data, and all the challenges met in this case would not have been possible without Bengt's open mind.

11.2   Mobile communication through the package

Case by Fredrik Nilsson

The fast-moving consumer electronic goods industry is known for its rapid change of products, with short product life cycles. While the supply and production of cell phones is similar for most actors worldwide, the sales channels differ and many are dependent on market structures and regulations. In the Scandinavian market, for example, the packaging is seldom shown until the product is purchased. Cell phones and digital cameras are usually sold in specialized stores where the products are displayed to the end customers, allowing them to physically touch and test them. In other markets and outlets, the packaging is the first thing customers see, but they are unable to touch and test the product before purchase. The packaging design can then be regarded as the silent salesman that communicates numerous aspects that can either promote or demote the product inside if the aspects are not fully considered and adapted accordingly.

Information on packages is crucial for all the supply chain actors, especially the consumers. This case was carried out at Sony Ericsson Mobile Communications AB and explores the role of packaging for cell phones in these two kinds of markets (Nilsson et al., 2013). This comparative case discusses packaging communication based on the supply chain impacts from a sustainability perspective. The markets, A and B, differ in many aspects concerning distribution and sales, but the packaging solutions are the same. In market A, consumers are only exposed to the product when it is sealed in its primary package in retail stores before they make a purchase. In market B, consumers are first exposed to dummy phones (non-activated display models of the actual phones) and do not see the primary package until the purchase is made.

11.2.1 The packaging system

The products of Sony Ericsson Mobile Communications are distributed and sold in many countries and at different outlets and distribution set-ups. The packaging system includes primary packaging solutions, the design of which is based on product requirements and brand image. The secondary and tertiary levels of packaging are usually designed based on pallet dimensions and/or the quantities that major customers order.

At large upmarket stores where the products are delivered by modern trucks in larger quantities, the consumer can test every cell phone model. But at small corner shops, where the products are delivered by small cars or even bikes, consumer promotion is only carried out by means of the sealed, primary packaging. In response to the trend to become more environmentally friendly, Sony Ericsson Mobile Communications has decreased the size of its primary packaging so that less material is used. This increases the fill rate during transport, but also impacts the package communication due to the smaller surface for printing information and other graphical items.

11.2.2 Packaging impacts on the markets

Market A is characterized by the retailers that dominate the supply chains. These are predominately smaller corner shops where the primary packaging functions as a marketing and sales tool. Most products cannot be tested by the end customer before purchase. The distribution network is under-developed and most of the distribution activities are carried out by smaller trucks, cars and bicycles. In market A, the primary packaging is always visible on the retailers' shelves and acts as the single-most important marketing and sales tool by its visibility there.

Market B is characterized by retail stores or outlets run by mobile network operators that dominate the market. Here, different kinds of dummy phones are most often used to promote the products. The primary packaging is never visible to the consumer before purchase. The consumers in this market can always feel a dummy before they purchase the actual product. The packaging is provided when the consumer has decided on his or her purchase and should match the consumer's perception of the quality of the product. For example, a fancy phone should have rather fancy packaging, while for those consumers who are environmentally concerned, the packaging should appear to be environmentally friendly.

When it comes to environmental solutions, there is a major difference between markets A and B in terms of primary packaging. One manager in market B stated: "We have to keep in mind that we always need to use a 'Green Heart' (environmentally friendly symbol) package that has paper and at least some other recycled material if at all possible." In market A, the consumer perception is different: The majority of the consumers think environmentally friendly packaging is of lower quality and looks cheap.

Figure 11.3 Difference in packaging size – Sony Ericsson vs. Nokia.

Thus, the retail actors in the supply chains of the two markets put forward the packaging in different ways. They also allow for the consumer to try the functionality of the product to varying degrees. Both these aspects affect the potential sales and the distribution of products.

In market A, the packaging communicates both with its shape and design and through the salesperson's pitch. In a number of interviews, the sales personnel reported that they used the information on the package when promoting the different alternatives for the consumers. This resulted in them being more interested and confident in selling a cell phone when they could explain all the functions of the phone rather than being unable to answer difficult questions from the consumers. In this market, the size of the package matters as one marketing manager expressed: "There should be a difference in our packaging portfolio. Model x and y, for example, have the same packaging but are in totally different price segments." The size also impacts the shelf space and particularly so when it comes to visibility for consumers (Figure 11.3). This is problematic in this case because the competitors have larger primary packages on which more information is provided.

11.2.3 Suggestions for primary packaging improvements

Based on the different supply chain requirements and the characteristics of the two markets, a number of improvement potentials where identified. Some are listed in Table 11.2.

11.2.4 Sustainable supply chain implications

This case presents some of the challenges that companies confront in doing a good job of making sales, while also trying to show that they are environmentally concerned. From a packaging perspective it shows the role packaging can play in enhancing the promotion of a product, but also in lowering the carbon footprint.

Table 11.2 Packaging improvement potentials identified on markets A and B.

Market A	Market B
• Provide more information about the features of the phone on the primary package. • Increase the package size according to price range of the products: he higher the price, the bigger the package. • Use shrink-wrapping to protect the primary packaging. • The number of primary packages per secondary packaging should be adjusted according to the price range of the products. • Increase the protection on the secondary packaging level.	• Evaluate possibilities for less expensive packaging material. • Use environmental initiatives and materials to enhance sales. • The number of primary packaging per secondary packaging should be adjusted according to the price range of the products. • Improved tape in order to reclose secondary packaging.

It was obvious in market A that packaging size matters for consumers and indirectly for the sales personnel, because it contains a good amount of information about the features; this gives the salesperson more confidence when promoting the cell phones. So, in order to sell, the packaging needs to communicate. This is especially the case in market A, where the information on the packaging communicates in two important ways: directly to the consumer when she/he looks at and reads the information on the package; and more importantly, to the salesperson, where the information helps her/him to promote and communicate the different features of the cell phones. This also means an increase in transports in which the larger packages contain more empty space.

The same product in market B can instead be packed in a smaller sized and an even more environmentally friendly package. In market B the need for more information on the outside of the packaging is reduced, since the consumer has a dummy phone to get the physical feeling and can get the information from the sales person. In markets like B, solutions with smaller packages having environmental appearance can contribute to improved sustainability.

In this comparative case the identification of local packaging requirements from consumers, sales personnel and sales contexts, as well as logistics and environmental aspects, illustrates the complex patterns between different market needs. From a sustainability perspective this is important to acknowledge, since the type and amount of information on the package can have an impact on sales, as well as on social and environmental issues (Table 11.3).

Case acknowledgements

Many thanks to Magnus Fagerlund and Jonas Körner for their contributions to the empirical work of the case.

Table 11.3 The impact (+ pros and – cons) on the three pillars of sustainability, from an environmentalley marked-adapted packaging (Market B).

	Planet	People	Profit
Producer	+ Less packaging material	– More packaging alternatives to handle	+ More products to the market – Increased packaging costs
Transport	+ Less empty space		+ Increased fill rate on pallets
Distribution Centre	+ Less package waste		
Distribution	+ Less empty space		+ Increased fill rate on pallets
Retail	+ Less package waste	+ Less product information on package	– Less sales based on information + Increase in sales to environmentally aware consumers
Consumer	+ Less packaging waste	– Less consumer information	+ More confident purchase decisions

11.3   Roll containers for dairy products: Connecting atoms and bits

Case by Daniel Hellström

Packaging is often the information link between the physical flow of packages in the supply chain and the IT systems that control that flow. By applying identification technology – such as barcodes, quick response (QR) codes, and radio frequency identification (RFID) – to packaging, you basically connect the bits of the information technology to the atoms of the packaging. This case illustrates the phenomenon. It also demonstrates that the information and communication functions of packaging have direct and indirect impacts on the three pillars of sustainable development.

The real-life case[3] was carried out at Arla Foods, which is one of the largest dairy companies in Europe that exclusively produces milk-based products. In contrast to most other retail suppliers in Sweden, Denmark and the UK, where the products are distributed by the retailers themselves, Arla Foods distributes fresh products directly to retail outlets. To do this efficiently, Arla Foods uses different types of returnable transport packaging (RTP) such as returnable pallets, reusable crates and trays.

---

3 Parts of this case were published in *International Journal of Physical Distribution and Logistics Management*, "The effect of asset visibility on managing returnable transport items" by Johansson and Hellström (2007), and by Hellström and Pålsson (2011) in *"RFID-Enabled Visibility in a Dairy Distribution Network"*, in *Intelligent Agrifood Chains and Networks*, pp. 267–280, Wiley-Blackwell Publishing Ltd.

In Sweden, Arla Foods has traditionally used one type of roll container in the distribution of dairy products. This roll container was specifically designed for use in the distribution of fresh milk, which constitutes the single greatest portion of the total volume distributed by the company. The roll container is used for external and internal transport at dairies and to display products in retail outlets. The introduction of new products and new primary packaging designs, however, has drastically increased the range of dairy products the company offers. This has affected the production and distribution processes. One example is the picking process that was previously performed by truck drivers at the truck platform and is now performed in a dairy distribution centre. The increased range of products has also resulted in greater volumes of the distributed products being placed and displayed on retail shelves and not in the traditional roll containers. To meet requirements regarding efficient picking and distribution of low-volume products, Arla Foods introduced a new roll container (Figure 11.4).

Like many other firms, Arla Foods has experienced difficulties in managing and keeping track of the RTPs. A large number are lost annually and information concerning how many RTPs are in circulation or how many are in stock at various points in the supply chain is not available. Based on historical purchases of RTPs, Arla Foods estimate that their shrinkage rate is approximately 10%, meaning that one out of ten are lost annually due to theft and misplacement. In total, Arla Foods must re-invest more than €2 million annually to cover the lost RTPs.

Figure 11.4 Arla Foods' new roll container.

Consequently, Arla Foods needed to come up with a better way to manage and locate the RTPs. The company implemented a tracking system for the 26,000 newly introduced roll containers at four dairy distribution centres. In total, the four distribution centres distribute approximately 500 stock-keeping units (SKUs) to 14,000 delivery points/retail outlets on a daily basis.

11.3.1 Implementing a tracking system

Arla Foods' dairies are made up of three physically integrated plants: a production plant; a warehouse for the finished products; and a dairy distribution centre. After production, the products are stored in the warehouse for finished products, and are distributed from there to Arla Foods' dairy distribution centres. In the dairy distribution centre, low-volume products are picked out and distributed to retail outlets. High-volume products are directly picked up from the warehouse for finished products.

Because Arla Foods needed to improve its control of the roll containers, the company implemented a tracking system with a unique identification for each one. To track the roll containers, Arla Foods used three identification points at each dairy: one in the receiving process; a second in the picking process; and a third at the repair shop (Figure 11.5). All roll containers pass through the first two locations, which enables Arla Foods to use two virtual zones for roll container localization: one internally in the dairy (i.e. roll containers that are received at the dairy); and another externally (i.e. roll containers that are in transport to the dairy or are at a specific customer's location). Thus, Arla Foods can receive information about exactly how many and which roll containers are located: 1) in the dairy; and 2) at a specific customer's, including those in transport to and from that customer. The third identification point at the repair shop was implemented to record the type of damage and repair needed for individual roll containers. For details about the tracking system and the implementation process, see Hellström (2009).

11.3.2 Implementation results

The implementation of the tracking system to control and manage roll containers has generally been successful for Arla Foods. The estimated risk of losing one in five roll containers annually was initially overcome by being able to track and trace and by taking action on potential shrinkage situations. As a result, hardly any roll containers were lost. Internal analyses at Arla Foods show that the elimination of lost roll containers at the beginning of the implementation was related to the fact that the people involved in the system became aware that Arla Foods was able to track and trace roll containers. Thus, the dairy organization, truck drivers and customers paid more attention to the rules and procedures. However, a few months after implementation, one dairy did no longer used its track-and-trace capabilities and instead fell back into old habits. This resulted in the shrinkage rate increasing to more than 15% of the roll container fleet on an annual basis. This highlighted the need for the tracking system and that tracking alone was not enough: proper actions and continuous attention from the management was also required to attain low shrinkage.

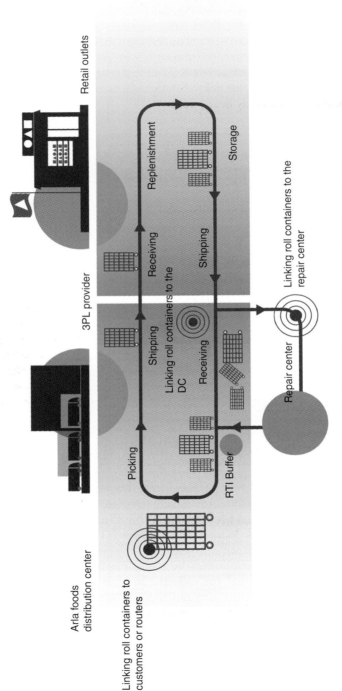

Figure 11.5 Rotation of the roll containers and the tracking identification locations. 3PL stands for third-party logistics.

In addition to reducing shrinkage, Arla Foods also gained other benefits. Based on the accumulated data, the company has been able to measure the rotation of the roll containers. One example is the distribution of the roll container cycle time, which indicates that nearly all roll containers are returned to the dairy within three to four days. This cycle time data, and data on how many roll containers are used in each customer shipment has enabled Arla Foods to simulate the rotation of the roll containers to determine the appropriate fleet size. This was done at one out of the four dairy distribution centres and indicated that the fleet size could be reduced by nearly 50%.

Accumulated data from the repair area have also been useful for Arla Foods. When roll containers are repaired, the individual identification number is registered and the type of damage and repair are recorded. This means that Arla Foods can identify the physical strengths and weaknesses of the roll containers and relate the damage to the producer, batch number, the previous customer/route of the roll container or its repair history. Based on this type of analysis, Arla Foods may be able to identify underlying reasons for problems resulting in damaged roll containers.

A rough return on the investment analysis indicated a payback period of approximately 14 months. The total cost of the tracking system was estimated at approximately €300,000. The running profit from the system is, however, difficult to verify. It is based on the assumptions that the investment would decrease the annual roll container loss by 7.5% (approximately €234,000 annually) and reduce the number of roll containers needed by 20% (approximately €31,000 annually). The tracking system resulted in nearly 0% loss of roll containers if appropriate actions were taken and continuous management attention was given, while a 20% annual loss was expected by Arla Foods to take place without the investment. This indicates that the running profit was based on very modest assumptions.

11.3.3   Concluding remarks: Connecting atoms and bits

The case shows that there are significant sustainability implications related to the information and communication functions of packaging (Table 11.4). Applying identification technology enables packaging to be used for accurate, reliable, timely and efficiently information gathering (De Jonge, 2004; McKerrow, 1996). With this information, logistical activities can be performed more effectively and with a lower environmental impact (Hellström 2009; Rosenau et al., 1996). Arla Foods comprehensively decreased its fleet size and shrinkage. The shrinkage level at Arla Foods is by no means unique. In a survey of 233 enterprises in consumer-oriented industries undertaken by the Aberdeen Group (2004), a quarter of the respondents reported that they lost more than 10% of their RTP fleet annually, with 10% of the respondents losing more than 15%. Finally, there is also a social impact. Since Arla Foods was able to track and trace roll containers, the dairy organization, truck drivers and customers paid more attention to the rules and procedures concerning the roll containers. This reduced their theft and misplacements.

Improving the information and communication functions of RTPs also has indirect impacts on sustainability. In the case of Aral Foods, these functions facilitated increased use of RTPs. The RTPs also provided Arla Foods with operational benefits such as improved protection and security of products, improved

Table 11.4 The impact (+ pros and – cons) on the three pillars of sustainability, of tracking and tracing roll containers at Arla Foods.

	Planet	People	Profit
Producer	+ Decreased fleet size (i.e. packaging material)		+ Decreased fleet size costs
			+ Decreased repair costs
	+ More efficient repair		+ Decreased reinvestment costs
	+ Less warehouse space needed		– Technology investment cost
Transport	+ Less shrinkage		
Distribution Centre			+ More accurate picking
Distribution	+ Less shrinkage	+ Less theft and misplacement	+ Less shrinkage
		+ Easier handling	
Retail	+ Less shrinkage	+ Less theft and misplacement	+ Less shrinkage
		+ Improved trustworthiness	

working environments, more efficient handling and fill rates, and reduced use of packaging materials and waste. Please note that these indirect benefits are not included in Table 11.4.

Case acknowledgements

Special thanks go to Eva Blomqvist and Berne Carlson at Arla Foods, Sweden. Access to the implementation process, all data, and all the field visits would not have been possible without these two wonderful people. Thanks!

11.4   What does the silent salesman do for a sustainable society?

Case by Annika Olsson

Groceries have been sold in self-service stores since the 1950s. Consumers walk around and make decisions of what to buy based on what they see on the shelves. The majority of the groceries sold are packed, and the package is the interface between the product and the consumer. But it is also the interface between the consumer and brand owners, who want to communicate the values of their

Figure 11.6 Package claiming that you get 33% extra, however, 33% or more is air.

products (Lindh et al., 2016; Olsson and Larsson, 2009). That is why primary packages for food are called the "silent salesman" (Löfgren and Witell, 2005).

But what does the communication from this silent salesman really mean to consumers for their decision-making in general, and for sustainable development in particular?

This case describes six real-life examples of communication, packaging information and design that can be found in stores and that are meant to affect consumer choices and decisions. The foremost aim of these messages is to attract consumers, but increasingly so, they also need to reflect what the product means for sustainable development and how it affects it. This case highlights and elaborates on sustainable implications based on the communication direction of the compass.

11.4.1 Dishwashing tablets

The role of the package as the silent salesman is to make products visible on the shelves in order to get the consumer's attention. One way of being visible is to be big! Another is the use of eye-catching colours and words, and the emphasizing of messages in bright colours. By promising that they will get more, this package for dishwashing tablets is graphically designed to emphasize the message that exclaims "33% extra", in bright yellow and red on the front and in large print. With such communication, consumers are led to believe that the package is brimming with the additional dishwashing tablets. However, in this example, illustrated in Figure 11.6, approximately 50% of the package contains the product and the rest is just air and empty space. Even though the package does indeed contain 33% more of the product (42 + 14 tablets) than the original, a half empty package results in a low fill rate and less efficient transports. This obviously affects the environment negatively and increases transport costs, while at the same time offering a better facing of the product on the retail shelves with a potential increase in sales. One risk with so much empty space in the package is consumer dissatisfaction when they open the box and wonder if the 33% extra might just be extra air.

11.4.2 *PlantBottle*

Environmentally concerned consumers focus on the packaging material when deciding what packages they think are environmentally friendly. Plastic packages are regarded by the majority of consumers to be more negative than paper ones made of renewable raw material (Lindh et al., 2016). The soft drink industry has met this criticism by introducing the *PlantBottle* (Figure 11.7) for water and carbonated soft drinks, signalling an environmentally friendly package with a green logotype that also has a recycling symbol. Further down on the bottle there is a clarification stating that up to 30% of the entire bottle is made from plants and that the entire bottle is 100% recyclable.

The material in these bottles is still made of 100% PET or PE, but up to 30% of the raw material used for making the PET/PE is based on ethanol made from plants. However, the origin of the plants is not described. The *PlantBottle*, consisting in part of bio-based polymers, is driving sustainable development in the sense that a renewable resource is used in the packaging material production. This is true as long as it is based on non-edible sources and thus does not "cannibalize" the food supply. Since the final packaging material is pure PE or

Figure 11.7 *PlantBottle* claiming that up to 30% is made from plants and is 100% recyclable.

pure PET, recycling in traditional bins for plastics is possible. The way this is conveyed with the green leaf and the word "plantbottle" may convince the consumer that the entire bottle is made with plants as the raw material. This may in turn make them more positive to buying the plastic bottle and, in this way, increase sales. From a sustainability point of view, the increase in the use of plant-based polymers is positive, as long as it does not affect the ability of the package to protect its content, and as long as the bio-based polymers are not cannibalizing the food supply.

11.4.3   Locally produced baby food

Environmentally concerned consumers prefer organic and locally produced food to a larger extent than more traditional consumers (Lindh et al., 2016). But it has not been clearly defined in the research or by the authorities what "locally produced" really means. Marketers, however, have certainly taken the opportunity to utilize consumers' environmental concerns in promoting their products. One of the main drivers for consumers to buy locally produced food is that they assume it has been transported shorter distances. The mango and banana baby food previously sold on the Swedish market is one example that communicated "locally produced" and "organic" on its label (Figure 11.8). The product was processed and packed in Swedish factories but the raw materials – the mangos and bananas – were obviously grown on another continent.

The negative effects on sustainable development may not be very large due to the distance, since transport usually constitutes only a small part of the environmental burden in food production, but it certainly raises questions in the heads of conscious consumers and less trust in the producers. Swedish consumers reacted

Figure 11.8 Label claiming the product to be organic and locally produced.

Figure 11.9 Package claiming organic cheese.

negatively to the statement "locally produced", since they know that neither of these fruits are native to Sweden. This resulted in the product being withdrawn from the market. The baby food jars also had the green "Krav"[4] label, which is a well-known seal of approval for organic food in Sweden. It is only a guarantee of the organic value of the product, and says nothing about the package or the transport.

11.4.4 Organic cheese packaging

In the same way, many brand owners emphasize the word "organic" in product names and use the package to signal these values. It is worth noting that the focus is exclusively on the product inside the package, while the package itself can have a lower level of sustainably. In this example, an organic cheese is double packed in a plastic jar covered by wrap-around paper (Figure 11.9). The plastic jar is sufficient to protect the cheese, but the paper is used to add the tactile feeling, as well as the mental feeling, that paper is a better material environmentally. Adding paper just for this purpose means adding even more to the negative environmental impact of the product/packaging system, since the plastic package will be enough for protective purposes.

11.4.5 Separable dairy package

One obstacle for consumers with the packaging for viscous food products such as yoghurts and sour cream has been in emptying the contents. This inability leads to unnecessary food waste. On average, 10–15% is wasted in packages that

---

4 Krav has been developing organic standards since 1985. The label stands for: sound, natural environment; solid care for animals; good health; and social responsibility. (www.krav.se/english), accessed 11 November 2015.

Figure 11.10 New separable package – "new smart packaging".

are thrown away, or rinsed down the drain from packages before recycling. A new solution to make it easier to entirely empty these containers has recently been launched as "new smarter packaging". The label tells the consumer that this package is separable and easier to empty because of a new feature that is good for the environment (Figure 11.10). The redesign with a "fully removable" top on the package is more expensive to produce because of the added operation of perforation, while still maintaining a tight seal. The idea behind making it more convenient for consumers to empty packages is good, and the added value to customers of being able to consume all of the yoghurt overrides the extra cost. But from a user-friendliness point of view, the new solution has proven to be less convenient, since the viscous product tends to spill and make things messy when the top is teared off. This leads to messy handling for consumers, and thus a risk that they do not use the feature aimed for reducing product waste. The question is then if the packaging producer can justify and benefit from the extra cost of the package, especially if it does not work satisfactorily from a user-friendliness point of view.

11.4.6 Three for the price of two

Another message that can result in food waste in silent sales communication is when consumers are urged to buy three packages for the price of two. Or when they are urged to buy more just because of a lower price compared to the price of buying it one by one. As in Figure 11.11, where consumers are asked to buy two packages of sausage for 15 Swedish kronor instead of one for 10.50 (about €1.5, instead of about €1 for one). This is sustainable only as long as the product has a shelf life that corresponds to the time it takes to consume the three. But if consumers take three, pay for two and throw away one, this negatively affects sustainable development.

Figure 11.11 Marketing saying buy more packages and save more.

References

Aberdeen Group (2004), *RFID-enabled Logistics Asset Management Benchmark Report*, Aberdeen Group, Boston, MA.

De Jonge P.S. (2004), *Making Waves: RFID Adoption in Returnable Packaging.* Logica CMG: www.can-trace.org/portals/0/docs/logicacmg_rfid_benchmark_study.pdf accessed February 2016.

European Commission, *Europa – Food Safty – Sustainibility of the Food Chain*: http://ec.europa.eu/food/food/sustainability/causes_en.htm

Hellström D. (2009), The cost and process of implementing RFID technology to manage and control returnable transport items. *International Journal of Logistics Research and Practice*, 12(1), 1–21.

Hellström D. and Pålsson H. (2011), RFID-enabled visibility in a dairy distribution network. In: *Intelligent Agrifood Chains and Networks*, Wiley-Blackwell Publishing Ltd, pp. 267–280.

Hernandez J. (2001), Food Safety. *Food Management*, 36(6), 84–86.

Johansson O. and Hellström D. (2007), The effect of asset visibility on managing returnable transport items. *International Journal of Physical Distribution and Logistics Management*, 37(10), 799–815.

Lindh H., Olsson A. and Williams H. (2016), Consumer perceptions of food packaging: Contributing to or counteracting environmentally sustainable development? *Packaging Technology and Science*, 29(1), 3–23.

Livsmedelsverket (National Food Agency, Sweden) (2004a), *Livsmedel transporteras vid för höga temperaturer – konsumenter vilseleds (Food transported at too high temperatures – consumers are misled)*: www.slv.se Accessed 3 April 2015.

Livsmedelsverket (National Food Agency, Sweden) (2004b), *Hur länge håller varan? (How long is the product good for?)*: www.slv.se Accessed 3 April 2015.

Loxbo H. (2011), *Hållbar konsumtion av jordbruksvaror. Matsvinn–ett slöseri med resurser*, Report.

Löfgren M. and Witell L. (2005), Kano's theory of attractive quality and packaging. *Quality Management Journal*, 12(3), 7–20.

McKerrow D. (1996), What makes reusable packaging systems work. *Logistics Information Management*, 9(4), 39–42.

Nilsson F., Fagerlund M. and Körner J. (2013), Globally standardized versus locally adapted packaging – A case study at Sony Ericsson Mobile Communications AB. *International Journal of Retail & Distribution Management*, 41(5), 396–414.

Olsson A. (2010), Value adding services in packaging – a value for all supply chain actors? In: *Conference Proceedings of 2nd CIRP IPS² Conference*, 14–15 April, Linköping, Sweden.

Olsson A. and Larsson A.C. (2009), Value creation in pss design through product and packaging innovation processes. Chapter 5. In: Sakao and Lindahl (eds), *Introduction to Product/Service-System Design*, Springer, pp. 93–108.

Rosenau W.V., Twede D., Mazzeo M.A. and Singh S.P. (1996), Returnable/reusable logistical packaging: A capital budgeting investment decision framework. *Journal of Business Logistics*, 17(2), 139–165. *waste basket*. In Swedish: *Rapport från en slaskhink*,

Stockholm Consumer Cooperative Society (2009), *Report from a* Konsumentföreningen Stockholm: http://www.konsumentforeningenstockholm.se

Stockholme Consumer Cooperative Society (2011), *Best-before and last day of consumption: What is the difference?* (In Swedish: *Bäst-före och sista förbrukningsdag på livsmedel. Vad är skillnaden?*, Konsumentföreningen Stockholm: http://www.konsumentforeningenstockholm.se/

WRAP (2007), *The food we waste*. Report: http://www.wrap.org.uk/

Afterword

The role of packaging in sustainable development is becoming a well-deserved high-profile issue. We argue that sustainable development is the essence of modern packaging design. It exists in so many forms and goes beyond single disciplines – meaning that all those who are involved or affected by packaging can promote sustainable development. Thus, no matter which development process, design methodology or tools we use, and no matter what organization or department we belong to – the all-encompassing and everlasting strategy of packaging design is to consider and contribute to sustainable development.

Equipped with the compass, we hope you will improve your abilities to manage and participate in packaging design projects and become more confident in designing packaging solutions that contribute to the three pillars of sustainable development. If you reach the point where the compass no longer brings much to the table, our goal with this book has been met. But our vision still remains: to advance packaging design knowledge in order to motivate and co-create holistic solutions for sustainable development.

To conclude, this book does not offer you the final formula on how to design packaging for sustainable development. It is rather a synthesis of the field and an attempt to develop a common language and understanding of what constitutes packaging design for sustainable development, and how this can be translated into practice. We clearly need more research on the topic, and it would be an understatement to say that much is still unknown about packaging design for sustainable development. It is somewhat alarming that the research on the topic is still in its infancy. This is certainty something that academia and industry jointly need to address in future R&D. Although this may be the end of the book, we are only at the beginning of something new. As ever, it will be interesting and exciting.

Managing Packaging Design for Sustainable Development: A Compass for Strategic Directions, First Edition. Daniel Hellström and Annika Olsson.
© 2017 John Wiley & Sons, Ltd. Published 2017 by John Wiley & Sons, Ltd.

Index

Page numbers in *italics* refer to figures; those in **bold type**, to tables.

Managing Packaging Design for Sustainable Development: A Compass for Strategic Directions,
First Edition. Daniel Hellström and Annika Olsson.
© 2017 John Wiley & Sons, Ltd. Published 2017 by John Wiley & Sons, Ltd.